Home Canning Meat, Poultry, Fish and Vegetables

Stanley Marianski
Adam Marianski

Bookmagic LLC,
Seminole, Florida

Home Canning Meat, Poultry, Fish and Vegetables
Stanley Marianski
Adam Marianski

ISBN: 978-0-9836973-7-4
Library of Congress Control Number: 2013940818

Bookmagic, LLC.
http://www.bookmagic.com

Printed in the United States of America.

CONTENTS

Introduction

The majority of books on canning cover the canning of fruit, jams and jellies using conventional processing in a water bath canner. These methods are usually passed from elder members of a family to younger ones, which can at times make the process scientifically incorrect. Canning low-acid foods such as meat or vegetables in an open canner is a dangerous procedure which must not be performed. Low-acid foods must be processed at different pressures and temperatures, a fact that is recognized by commercial packers. They are well aware of the potential risks involved and work under strict guidelines and continuous supervision by meat inspectors. When those guidelines are followed it is unlikely that a contaminated product will be produced in a factory. In most cases a defective container is the result of human error.

Home canned products do not fall under the jurisdiction of government agencies so the rules are not enforced. This explains why home canned products account for the majority of food poisoning cases. A hobbyist is at greater risk due to insufficient information on the subject and very few professionally designed and tested recipes.

It is especially difficult to find any information on canning food in metal cans, which is strange if one considers that most canned foods in the world are packed in metal cans. Obviously, if this technology was inadequate, commercial manufacturers would not be allowed to use metal cans for canning foods.

Home Canning of Meat, Poultry, Fish and Vegetables explains in simple language the science of canning low-acid foods and reveals the procedures that are used by the canning industry. The material is based on the U.S. government requirements as specified in the Code of Federal Regulations and the relevant links are listed. After studying the book, a newcomer to the art of canning will be able to safely process foods at home in both glass and metal containers.

Stanley Marianski

Chapter 1

Preserving Food by Canning

The purpose of canning is to use heat, often with other means of preservation, to kill or inactivate all microorganisms and to package the product in hermetically sealed containers so that it will be protected from recontamination.

Meats can be preserved for a long time without any loss in quality, ordinary temperatures have no effect on them. Fats or sausages become rancid in time even when kept in a freezer. Canned meats are less prone to such problems. Canned meats are immediately ready to serve and can be taken on outdoor trips, contrary to frozen foods that have to be thawed out first. Animals and insects cannot force their way through a can or the jar, so the safeguarding of food is easier. Pork, beef, poultry, fish, wild game and dishes that contain those meats can be successfully canned in glass jars or metal cans and stored for up to one year for best quality. They will still be edible in 2-3 years. Either fresh meat or frozen meat may be used for pressure canning. Up north where hunting is a part of daily life, many Alaskans freeze their game meat for up to one year. Then the meat is preserved using the pressure canning process. This gives meat an effective shelf life of two years.

History

Napoleon used to say that the army marches on its stomach. The staple of a soldier was salted meat, stale bread and whatever fresh food he could pick up as he moved along. The wars had an effect on the common people as well and food shortages were rampant. The government strongly believed that the condition of both the army and the civilian population would be greatly improved if only a better method of preserving food would be developed. To this effect, in 1795 the French military offered an award of 12,000 francs to a person who could come up with a practical solution.

Nicolas Appert was a confectioner and chef in Paris from 1784 to 1795. He had great skill with food and was also familiar with brewing and distillation. He set up a small kitchen in the back of his shop and decided to answer the challenge. For ten years he labored patiently with different foods, cooking them in different ways and trying to preserve them. There was no scientific papers that Nicolas Appert could draw upon in his food experiments as bacteria would be discovered by Louis Pasteur fifty years later.

P1.1 Nicolas Appert.

Drawing by Tadeusz Kasperkowicz.

After many experiments Appert concluded that air was the cause of food spoiling and he tried to remove as much of it as possible. Neither glass jars nor metal cans were around yet and Nicolas Appert had to use bottles. Those being narrow were ill suited to fill with food so Nicolas designed his own wide mouth bottles, which looked like glass bottles that were used for daily milk deliveries later on. A cork was used to seal the bottle. It took Appert fifteen years to perfect his method and after many experiments he arrived at the correct times for preserving many foods. The method he developed does not vary from our present canning techniques, the difference being mainly in equipment and a better understanding of the process that we possess today. The method that Appert developed involved:

- Pre-cooking the food.
- Bottling it.
- Wiring the corks to prevent them from shooting out under pressure.
- Placing bottles in burlap sacks.
- Cooking bottles in a big kettle.

After fifteen years of experimenting and sending his canned products all over the world for testing, Appert presented his notes to the French government. On January 30, 1810, the Minister of Interior informed Appert that after careful examination of his process the prize money is awarded to him. His findings were compiled in a little book called *L'Art De Conserver - The Book for All Households or the Art of Preserving Animal and Vegetable Substances for Many Years.* A new word "appertization" became synonymous with processing foods using a hot water bath. In his process Appert used large quantities of sugar, salt and vinegar as preserving agents.

It should be emphasized that although Appert has invented the canning method, the cause of spoilage of food was still a mystery. Canning was more of a curiosity topic for scientists than a practical invention for the general public. Many scientists of the day tried to explain his process without much success. The best conclusion was that air is combined with food in a sealed bottle and in some mysterious fashion this magical combination prevented the spoiling of foods. It was universally agreed that the process worked and the ignorance apart, this canning method was employed for the next 50 years. In the coming years both the understanding of the process and better equipment led to the ever increasing popularity of canning.

The Timetable of Canning History

1795 - French government offers award for a new method of food preservation.

1810 - Nicolas Appert invents canning method and wins the award and prize money.

1810 - Peter Durand of England, receives patents for glass and metal canning containers.

1820 - First canning plants start to operate in Boston and New York.

1823 - A can with a hole in the top is invented.

1824 - Nicolas Appert develops a schedule for processing 50 different foods.

1840 - The tin container is in widespread use in the US.

1851 - Chevalier Appert (son of Nicolas Appert) develops the pressure cooker.

1860 - Louis Pasteur discovers microorganisms.

1864 - First Pacific salmon is canned.

1870 - About 100 canning plants are operating in the USA.

1874 - An American, A.K. Shriver of Baltimore invents commercial steam pressure retort.

1883 - The Norton Brothers Company of Chicago invents semi-automatic can soldering equipment.

1890 - Prescott and Underwood from Maine canneries, Russell in Wisconsin, and Barlow in Illinois discovered the relationship between thermophillic bacteria and the spoilage of canned food.

1895-1900 - The science of bacteriology is applied to canning.

1899 - Campbell produces the first canned condensed soups.

1901 - Norton Brothers merges with 60 other firms to form the American Can Company (123 factories).

1906 - The Chemical Laboratory of American Can Company is established.

1909 - Tuna fish is first canned.

1910 - The basic biological and toxicological properties of *Clostridium botulinum* are discovered.

1920's - Bigelow and Esty established the relationship between the acidity (pH) of foods and the heat resistance of bacterial spores. This determination laid the foundation for the classification of canned foods into acidic foods and low-acid foods.

1920 - Citrus fruit and tomato juice are canned.

1920 - Charles Olin Ball (1893-1970) develops thermal death time studies which become the standard for the United States Food and Drug Administration (FDA) for calculating thermal processes in canning. After earning his PhD from George Washington University in 1922, Ball would work with the American Can Company in Illinois and New York where he earned 29 patents. He would work at Owens-Illinois Glass Company during 1944–1946 before joining Rutgers University as a professor and later chair of the food science department from 1949–1963.

1940 - Canning plants operate all throughout the US.

In the US, canning became popular due to its shelf life, but in Europe it was a necessary survival skill. Many countries throughout history were involved in continuous wars which left buildings, stores and infrastructures destroyed. After the second world war ended in countries such as Poland or Germany many cities were 80% destroyed. People could not buy food in stores because there were no stores. What was left was rubble and piles of bricks. All citizens could do was buy food at farmers' markets and preserve it by canning for later use. If items such as tomato paste, wine, sauerkraut, pickles, pickled mushrooms, canned meats, or lard were not processed in the summer, there would be famine in the winter time. As refrigerators were not common, the food was canned and stored in root cellars or kitchen pantries.

Development of Canning in the US

On May 15, 1862, Abraham Lincoln signed into law the Agricultural Act that established the U.S. Department of Agriculture. The department's primary focus was to stimulate food production by providing seed and agricultural information to farmers and help them receive a fair price for their crops.

Following the U.S. Civil War, westward expansion and development of refrigerated railroad cars spurred the growth of not only the livestock industry, but also meat packing and international trade. In response to the growing pressure from veterinarians, ranchers, and meat packers to eradicate livestock diseases in the United States, President Chester Arthur signed the Bureau of Animal Industry Act. The Act created the USDA's Bureau of Animal Industry (BAI) in 1884, effectively the true forerunner of Food Safety and Inspection Service (FSIS). In 1905, the BAI faced its first challenge with the publication of Upton Sinclair's *The Jungle*. The ground breaking book exposed unsanitary conditions in the Chicago meat packing industry, igniting public outrage, which eventually led to the establishment of continuous governmental inspection.

Compared to salting fish, smoking and drying meats, or making salami type dried sausages, canning was a relatively new method of preserving food. Although small canning plants appeared in different areas of the country, the food was canned in all types of containers. In 1901 canning becomes a big industry when the Norton Brothers merged with 60 other firms to form the American Can Company (operating 123 factories). In this emerging period, losses from spoilage, as well as from poor quality were accepted as normal. Refrigerators were not common yet and the convenience of storing food at room temperature far outweighed a little loss of quality. There was, however, a big problem. People got sick and often died, a fact that was not unnoticed and had to be dealt with.

In response to both *The Jungle* and the Neill-Reynolds report, Congress passed the Federal Meat Inspection Act in June 1906. The Act allowed the USDA to issue grants of inspection and monitor slaughter and processing operations, enabling the Department to enforce food safety regulatory requirements. In 1910, the Meat Inspection Division established a research center in Beltsville, Maryland. Seven similar laboratories were later created throughout the country. These laboratories were responsible for both developing new testing methods and testing meat and meat products for foreign substances. A big problem was a lack of reliable formulations. In 1878 *The Canning Trade* magazine was created and it started publishing technical articles and the first canning formulas. In 1914 the best material was combined into one publication and the first edition of *A Complete Course in Canning* was printed. In 1996 the 13th edition of *A complete Course in Canning* was published. The work, still published by Canning Trade Inc., has become the textbook for students, packing plants and anybody employed in the canning industry.

Farmer's Bulletin 359

The earliest USDA publication for home canning was the *Farmer's Bulletin 359*, issued in 1909 by the Bureau of Chemistry. A discussion of decay as caused by molds, yeasts and bacteria was included along with an explanation that air must be excluded not for its own damaging properties but to exclude bacteria. It was explained that proper sterilization required heat. The process recommended was fractional sterilization - "the whole secret of canning." It recommended heating the vegetable in the jar to the boiling point of water and maintain that temperature for 1 hour each for two or three successive days. The first day of boiling was to kill molds and almost all the bacteria, but not spores. The spores were thought to germinate upon cooling, and boiling the second and third days killed the new bacteria. If fractional sterilization was not practiced, about 5 hours of boiling on the first day was recommended.

Farmer's Bulletin 839

This was the next publication issued in 1917 by the States Relation Service which later became the Extension Service, *"Home Canning by the One-Period Cold-Pack Method."*

Three basic processes were recommended for fruits, vegetables and meats: boiling water bath (212° F), water seal process (214° F), and steam pressure process. Screw top containers were recommended. This bulletin devoted much of the space to operating the canning equipment and less to the theory behind sterilization and spoilage, a month later *Home Canning of Fruits and Vegetables* is issued.

Farmer's Bulletin 853

Home Canning of Fruits and Vegetables was issued in 1917. It explained the causes of spoilage such as exposure to molds, yeast, bacteria, spores, and enzymes. A distinction was made between sterilization (the killing of all microorganisms) and pasteurization (heat treatment which kills vegetative cells but not spores).

Farmer's Bulletin 1211

Both *Farmer's Bulletins 839* and *853* were replaced in 1921 with *Home Canning of Fruits and Vegetables*, Bulletin 1211 (USDA 1921). The publication covered in detail the "whys" of processing and reasons for spoilage.

Canning Theory Development (1920-1925)

During those first experimental years there were numerous outbreaks of botulism caused by commercially canned products. To combat the growing problem, in 1923 the USDA establishes Bureau of Home Economics where many experimental works on canning were performed. The recommendations are issued for canning vegetables (except tomatoes) in pressure canners and water bath canners for fruits and tomatoes. Timetables were given for both pint and quart jars and tin cans.

Until 1920 thermal processes were based on individual experiences rather than on scientific knowledge. The science of bacteriology was not yet developed enough to answer the needs of the growing canning industry. There was a growing library of literature on the thermal death point of microorganisms and spores of *Cl. botulinum*. Results of the research showed that the heat resistance of microorganisms was affected by pH levels, age of the spores, and sodium chloride. This was a learning period and not all statements were true, for example USDA argued that the spores of *Cl. botulinum* were destroyed by heating for one hour at 175° F, 80° Celsius. Burke (1919), however, concluded in her experiments that spores of *Cl. botulinum* will survive boiling for 3.5 hours. This indicates kettle canning is not a reliable method to sterilize material contaminated with bacterial spores. In 1921 Weiss showed resistance of *Cl. botulinum* spores to 212° F (100° C) for up to 5 hours and he demonstrated that the thermal death point of *Cl. botulinum* could be affected by temperature and time of exposure, syrup density, food consistency, and acidity of the food. As more proof became available, the USDA started to incorporate new knowledge into its home canning publications.

In 1923 Charles Ball developed thermal death time formulas which relied less on empirical data. His formulas could be adapted to all can sizes and retort temperatures. His research become the standard for the United States Food and Drug Administration for calculating thermal processes in canning.

Adaptation Period (1926-1946)

Not much canning research was done in 1926-1939, however, between 1930-1935 plenty of studies were done on beef, chicken and other meats. The findings were incorporated as new recommendations into new publications.

National Canners Association Bulletin L-26

On January 18, 1930, the Board of Directors of the National Canners Association (now National Food Processors Association) approved for publication process suggestions for various *low-acid* foods packed in *metal* containers. The first edition of the Bulletin was published the same month. Whenever further data and information have become available, the Bulletin has been revised. *Bulletin 26-L* has proven itself to be extremely popular with the canning industry and the 13th edition was published in 1996.

Farmer's Bulletin 1762

In 1936 a new bulletin for canning fruits, vegetables and meats was issued. The processing times were calculated for 15 PSI pressure (250° F, 121° C) and were provided for ½ pint and quart sized jars, and No. 2, No. 2.5 and No. 3 cans. Those new timetables for meat were in effect from 1935-1942. Those recommendations were for the most part in excess of the required processing as the 15 PSI pressure (250° F, 121° C) was recommended.

Conservation Bulletin 28

In 1942, the U.S. Department of the Interior issued *Conservation Bulletin 28, "Home Canning Fishery Products."* It was strongly emphasized that "under no circumstances should any fishery product be canned unless a pressure canner is used. It is impossible to obtain a sufficient heat treatment or process, by any other means."

AWI Publications

In 1943 USDA issues *AWI-61*, "Canning Tomatoes", and *AWI-41*, "Wartime Canning of Fruits and Vegatables" which superseded *Farmers Bulletin 1762*.

In 1944, *AWI-93, "Home Canning of Fruits and Vegetables"* replaced *AWI-41* and *AWI-61*. Oven canning was labelled "dangerous" due to ineffective and serious accidents. Open kettle canning was labelled as "wasteful" for fruits and tomatoes, "dangerous" for vegetables and was suggested only for preserves, pickles or other foods with enough sugar or vinegar to prevent spoilage. The timetable for fruits offered boiling water bath processes' for vegetables, pressure processes at 240° F, 116° C were recommended.

The majority of current USDA process recommendations for low-acid foods (vegetables, meats, poultry, fish) are the result of three years of extensive research between 1944-1946. The earlier guides, for example an excellent *Montana Extension Service No. 242* bulletin "Canning Meat, Fish and Poultry" listed extremely safe procedures that called for long processing times at 250° F, 121° Celsius.

It was later deemed unnecessary and 240° F, 116° C, temperatures were found satisfactory. The final findings were published in USDA bulletins:

AWI-110, USDA, 1945, *"Home Canning of Meat."* Processing instructions were given for canning in glass jars *and* metal cans.

AIS-64, USDA, 1947, *"Home Canning of Fruits and Vegetables."*

Home and Garden Bulletin No. 8, USDA, 1947, "Home Canning of Fruits and Vegetables."

A description of the research and the data were released in 1946 in Technical Bulletin No. 930 - *"Home Canning Processes for Low-Acid Foods."*

The latest *USDA bulletin 539*, Dec 2009, *Complete Guide to Home Canning* still lists the original processing times for low-acid foods canned in glass jars in 1946. The USDA tables for determining proper process times include processing times with altitude adjusted for each product. Process times for ½ pint and pint jars are the same as times for 1½ pint and quart jars in the 1946 guide. For some products you have a choice of processing at 5, 10, or 15 PSI. In these cases, choose the canner pressure you wish to use and match it with your pack style (raw or hot) and jar size to find the correct process time. These guides should be studied as they are the best reference material for a hobbyist.

The National Canners Association was involved in establishing processing times and temperatures in *USDA AWI-110, 1945, "Home Canning of Meat."* The National Canners Association published its own bulletins:

Bulletin 26-L, Thermal Processes for Low-Acid Foods in Metal Containers, 1930, 13th edition was published in 1966.

Bulletin 30-L, Thermal processes for Low-Acid Foods in Glass Containers, 6th edition was published in 1991.

The National Canners Association has changed its name to the National Food Processors Association (NFPA) and in January 2005 it became the Food Products Association (FPA). Commercial packers that thermally process shelf-stable products in hermetically sealed containers must comply with the Food Safety and Inspection Service

(FSIS) regulations on the canning of meat and poultry products - Title 9 of the Code of Federal Regulations: Part 318, Subpart G, and Part 381, Subpart X. The Good Manufacturing Practices (GMP) regulations in the 21 CFR 108, 113, and 114 became effective May 15, 1979. These regulations are designed to prevent public health problems in low-acid and acidified low-acid canned foods. Low-acid canned animal foods are also regulated by the FDA with similar regulations and training requirements in the 21 CFR 507 and 508. Similar regulations in 9 CFR 318.300 and 381.300 regarding processed meat and poultry products were implemented by the USDA and FSIS on June 19, 1987. FSIS requirements for training supervisors of thermal processing and container closure operations became effective on December 19, 1988.

Better Process Control Schools

In 1963, 1971, 1978 and 1982 there were botulism incidents in commercially produced cans in the US. The 1971 incident led the National Canners Association (NCA), now Grocery Manufacturers Association/Food Processors Association (GMA/FPA), to reevaluate the commercial sterilization processes for low-acid foods. The result was a recommendation to the U.S. Food and Drug Administration for a program known as the NCA-FDA Better Process Control Plan. The plan became effective in 1973 and is known today as 21 CFR Part 113 - Thermally Processed Low-Acid Foods Packaged in Hermetically Sealed Containers.

The Better Process Control Plan (BPCP) places responsibility for the production of safe food products on individual food industry employees. The program stresses the point that " in the processing of food, no amount of mechanical devices, regulations, inspections or physical measurements can prevent or offset human error and the resultant potential for tragic consequences to consumers, processors or the industry in general."

The BPC Plan requires that operators of thermal processing and packaging systems work under the supervision of a person who attended, completed and passed the exam of a prescribed course of instruction at a school approved by the FDA Commissioner. The 4-day course is a cooperative venture of universities, FDA and industry personnel.

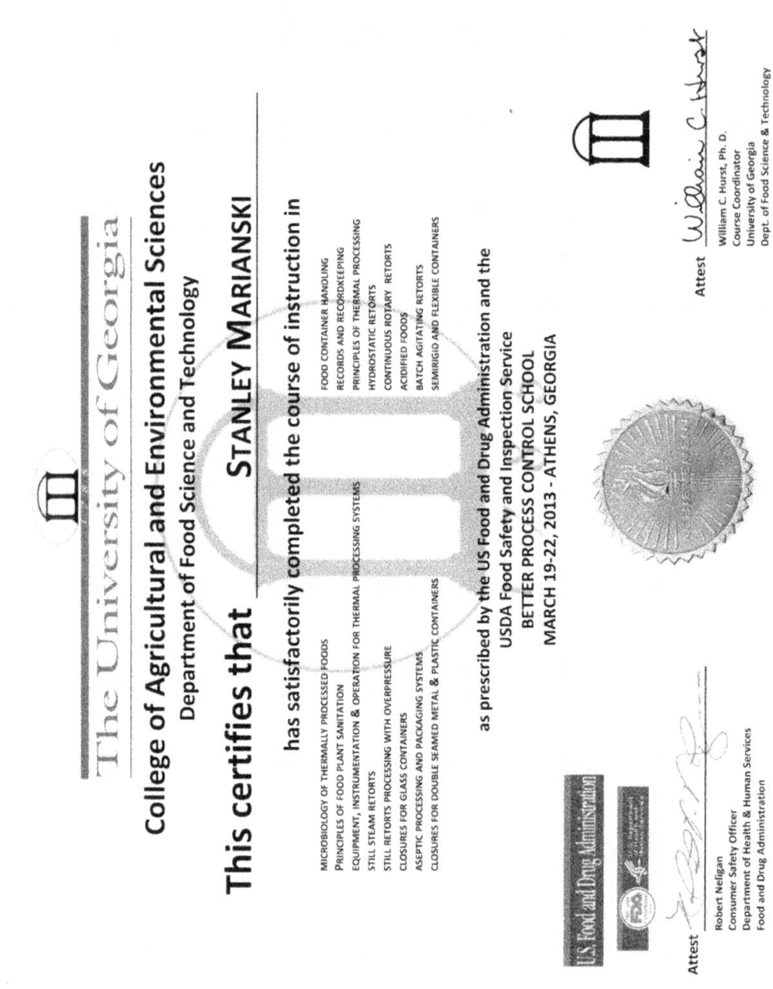

The University of Georgia

College of Agricultural and Environmental Sciences
Department of Food Science and Technology

This certifies that

STANLEY MARIANSKI

has satisfactorily completed the course of instruction in

MICROBIOLOGY OF THERMALLY PROCESSED FOODS
PRINCIPLES OF FOOD PLANT SANITATION
EQUIPMENT, INSTRUMENTATION & OPERATION FOR THERMAL PROCESSING SYSTEMS
STILL STEAM RETORTS
STILL RETORTS PROCESSING WITH OVERPRESSURE
CLOSURES FOR GLASS CONTAINERS
ASEPTIC PROCESSING AND PACKAGING SYSTEMS
CLOSURES FOR DOUBLE SEAMED METAL & PLASTIC CONTAINERS

FOOD CONTAINER HANDLING
RECORDS AND RECORDKEEPING
PRINCIPLES OF THERMAL PROCESSING
HYDROSTATIC RETORTS
CONTINUOUS ROTARY RETORTS
ACIDIFIED FOODS
BATCH AGITATING RETORTS
SEMIRIGID AND FLEXIBLE CONTAINERS

as prescribed by the US Food and Drug Administration and the
USDA Food Safety and Inspection Service
BETTER PROCESS CONTROL SCHOOL
MARCH 19-22, 2013 - ATHENS, GEORGIA

Attest Robert Neligan
Consumer Safety Officer
Department of Health & Human Services
Food and Drug Administration

US Food and Drug Administration

Attest William C. Hurst, Ph. D.
Course Coordinator
University of Georgia
Dept. of Food Science & Technology

Photo 1.2 Better Process Control School Certificate. This top quality canning course is sanctioned by the FDA and USDA/FSIS and conforms to all government regulations.

US Regulations For Canned Foods

Neither the FDA nor the USDA/FSIS have jurisdiction over foods that are canned at home. This creates big safety problems for low acid-products such as meat, poultry, fish and vegetables that are produced at home. Too often, home canners who have produced jams in the past incorrectly assume that all foods can be processed with the same procedure. This may in part be attributed to the insufficient information on the subject of canning low-acid foods that has been written for a home canner, but ultimately, the responsibility rests with the person who makes the product. He cannot blame his ignorance on the canning techniques of his mother or grandmother, as they did not have access to the information that is in abundance today. *It is his duty to learn the basic rules of canning before he attempts to make a product.*

Almost all cases of food poisoning can be traced to canned food produced at home. It is very rare that a commercial plant will employ incorrect procedures or use a faulty recipe. The commercial packers work under such tight regulations and inspection programs, that any violation of the process is usually the product of a human error.

Commercial canning is controlled in the United States by the Food and Drug Administration and the United States Department of Agriculture's Food Safety Inspection Service. We enclose below the list of canning regulations that must be followed by commercial packing plants. Although home canning is not subject to government control, we feel that anybody who is seriously interested in proper canning procedures will greatly benefit from this information.

Government regulations are the result of 100 years if intensive studies and research. The FDA and the USDA are concerned with the safety of consumers and not the taste or flavor of the canned product. However, each formulation must be designed by a competent processing authority and approved by the FDA. The regulations listed in the Code of Federal Regulations will help anybody who is thinking of starting his own commercial packing venture.

Code of Federal Regulations may be accessed for free online at:

ELECTRONIC CODE OF FEDERAL REGULATIONS

http://www.ecfr.gov/cgi-bin/ECFR?page=browse

Food and Drug Administration (FDA)

Title 21, Code of Federal Regulations (**21 CFR**)

Part 108 - Emergency Permit Control.

Forms required for food canning establishment registration and process filing with FDA:

FD-2541 Registration for Canning Establishment.
FD-2541a Process filing for all processing methods, except aseptic processing of low-acid foods.
FD-2541c Process filing for aseptic processing of low-acid foods.
Forms can be downloaded from:
http://www.fda.gov/AboutFDA/ReportsManualsForms/Forms/default.htm

Part 113 - Thermally Processes Low-Acid Foods Packed in Hermetically Sealed Containers.

Part 114 - Acidified Foods.

Fish products fall under FDA jurisdiction.

United States Department of Agriculture (USDA) - Food Safety Inspection Service

Note that meat and poultry products fall under the jurisdiction of the USDA/FSIS. This should not create problems or confusion since both the FDA and USDA/FSIS regulations are almost identical.

Title 9, Code of Federal Regulations (9 CFR)

Subpart G - CANNING AND CANNED PRODUCTS

Part 318.300 - .311 Meat Production.
Part 381.300 - .311 Poultry Production.
Part 416 - Sanitation Procedures, Meat and Poultry.

Useful General Information

Part 110 - Good Manufacturing Practices in the manufacture, packing and holding of food.

Part 120 - Sanitation Procedures, Juice and Juice Products.
Part 123 - Sanitation Procedures, Fish and Fishery Products.

According to the FDA regulation 21 CFR Part 113, a canned food with a water activity greater than 0.85, and a pH greater than 4.6 is considered a low-acid food, and its heat process will have to be filed by the individual packer with the *FDA*. If reduced water activity is used as an adjunct to the process, the maximum water activity must also be specified.

If the pH of the product has been adjusted to 4.6 or less and the water activity is greater than 0.85, the product is covered by the acidified food regulation, 21 CFR, Part 114, and requires only enough heat to destroy vegetative bacterial cells.

All low-acid and acidified low acid meat or poultry products are subject to the USDA canning regulations if the water activity is greater than 0.85.

Any non-meat food product, regardless of the pH, with a water activity of 0.85 or less is not covered by the regulations for either the low-acid food (21 CFR, Part 113) or the acidified food (21 CFR. Part 114). However, these products are covered by the FDA's Current Good manufacturing Practices (CGMPs) regulation 21 CFR, Part 110.

Meat or poultry products with a water activity of 0.85 or less are not covered by the USDA/FSIS canning regulations but are covered by other regulations, such as the USDA/FSIS meat and poultry Hazard Analysis and Critical Control Points (HACCP) regulation (9 CFR, Part 117) and Sanitation SOP regulation (9 CFR, Part 416).

A Complete Course in Canning

There is a book or rather "work" that has been perfected for over one hundred years and in our opinion it stands above others. This work is *A Complete Course in Canning* that has been published by The Canning Trade Inc., Baltimore, Maryland since the 1900's. What has started as a single book, has become in time a two volume set and then the three volume set of books. Of particular interests are the 9-12 (1970-1987) editions which were revised and enlarged by Anthony Lopez, Ph.D., professor of food science and technology at Virginia Polytechnic Institute and State University at Blacksburg, Virginia and anything published later. The latest 13th edition (1996) has been edited by Donald L. Downing, Ph.D., from the New York State Agricultural Experiment Station located in Cornell University, Geneva, New York.

This monumental work is a technical reference and textbook for students of food technology, plant managers, product research specialists, food equipment manufacturers and everybody who is professionally engaged in the canning trade. The complete set totals almost 2,000 pages and might be overkill for the average person, but it is of an immense value to a person that wants to master all aspects of canning or is contemplating his/her own canning business.

Chapter 2

Glass Jars and Metal Cans

When Nicolas Appert struggled with his first attempts at canning, neither glass jars nor tin cans were around, so he used wide mouth bottles and corks which were secured with wire. John Mason patented the continuous-thread closure on November 30, 1858. In 1883 William Charles Ball and his brothers switched over from tin containers to glass containers for storing lard and oil. In 1886 they started to use glass containers for storing fruits. In 1903 Alexander Kerr started to use wide mouth glass jars. Later in 1915 he invented a smaller, flat metal disk with a permanent composition gasket. His two-part lid system transformed home canning safety and is still in use today.

Glass jar is made from three separate molds: finish, body and bottom, which are then molded together. The connection is known as mold seam and may be mistaken for a crack in the glass. Each part has its own mold seam and they may or may not be aligned in a straight line.

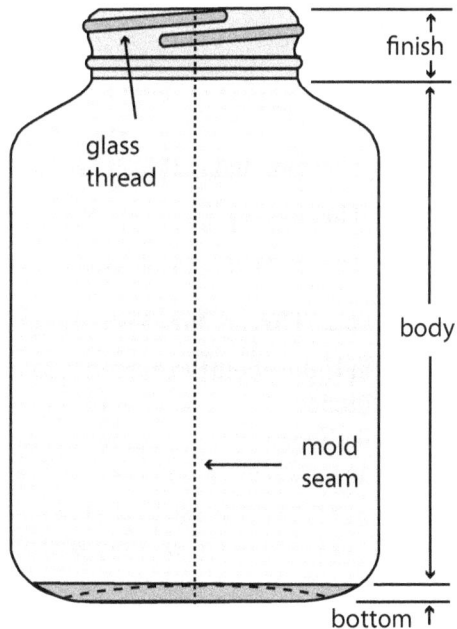

Fig. 2.1 The basic parts of a glass jar.

Use only standard jars intended for home canning. Never use jars from commercial food products. Regular and wide-mouth Mason-type, threaded, home-canning jars with self-sealing lids are the best choice. They are available in ½ pint, pint, 1½ pint, quart, and ½ gallon sizes. The standard jar mouth opening is about 2⅜ inches. Wide-mouth jars have openings of about 3 inches making them more easily filled and emptied. Half-gallon jars may be used for canning high acid juices. Regular-mouth decorator jelly jars are available in 8 and 12 ounce sizes. With careful use and handling, Mason jars may be reused many times, requiring only new lids each time. When jars and lids are used properly, jar seals and vacuums are excellent and jar breakage is rare.

Most commercial pint and quart size mayonnaise or salad dressing jars may be used with new two piece lids for canning acid foods. However, you should expect *more seal failures and jar breakage.* Seemingly insignificant scratches in glass may cause cracking and breakage while processing jars in a canner. *Mayonnaise-type jars are not recommended for use with foods to be processed in a pressure canner because of excessive jar breakage.* Other commercial jars with mouths that cannot be sealed with two-piece canning lids are not recommended for use in canning any food at home.

Glass Jar Types

There are three types of glass jars and closures currently used in food packaging:

- The continuous-thread jar.
- The twist-on jar.
- The press-and-twist jar.

Continuous-Thread Jar With Single Lid

The jar and lid are both threaded in one continuous bead around the entire circumference of the opening, referred to also as a "screw-on". There is less likelihood of over-tightening a screw-on closure. Continuous-thread jar may have a one part lid, for example mayonnaise jar or it may have a closure consisting of two parts: a lid and a screwband. An example is your typical canning Mason jar.

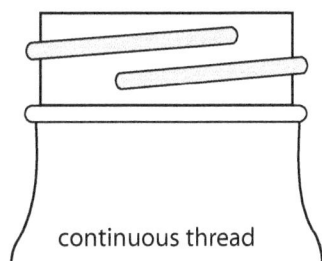

Fig. 2.2 Continuous thread jar.

Photo 2.1. Continuous thread jar.

Photo 2.2 Continuous thread lid.

Photo 2.3 Continuous thread lid. The lid is characterized by gripping ribs on the outside.

All types of jars can have a vacuum indication button built into the lid.

Many believe that a canning jar must be embossed on the sides with the word Mason, Ball or Kerr. Well, what is the real Mason jar? Any jar with a continuous-thread closure will do the same job as long as it is made strong enough to withstand the thermal shocks encountered in the canning process.

Continuous-Thread Jar With Two-Piece Lid

The advantage of glass jars with two-part lids lies in the fact that those jars have been around in the USA for so long that they have become an unwritten standard in home canning. There are three sizes available: ½ pint, pint and quart, two lid sizes will fit all jars, and the jars can have a regular or wide mouth opening. The lids come with one universal sealing compound and the jars have been perfected over the years. If used properly they create a perfect seal every time.

Photo 2.5 Pint and ½ pint jars above.

Photo 2.4 Quart size jars. Wide mouth-left, regular-right. The capacity of both jars are equal, but the diameter of the lids are different.

It is our recommendation to use wide mouth jars for canning meat as many products will hold their shape better when removed from a jar, for example meat loaf. Then they can be browned or roasted and they will look much better on a plate. There are meats that are eaten cold, for example head cheese or meat jelly. You can use regular jars for soups, beef stews or products which are made of many smaller components, for example chili con carne and beans. Canning is simpler, however, when only wide mouth openings are chosen. In addition, the same diameter lid and screwband fit both, pint and quart size jars.

Lid Selection, Preparation, and Use

The common self-sealing lid consists of a flat metal lid held in place by a metal screw band during processing. The flat lid is crimped around its bottom edge to form a trough, which is *filled with a colored gasket compound.* When jars are processed, the lid gasket softens and flows slightly to cover the jar-sealing surface, yet allows air to escape from the jar. The gasket then forms an airtight seal as the jar cools. Gaskets in unused lids work well for at least 5 years from date of manufacture. The gasket compound in older unused lids may fail to seal on jars.

Buy only the quantity of lids you will use in a year. To ensure a good seal, carefully follow the manufacturer's directions in preparing lids for use. Examine all metal lids carefully. Do not use old, dented, or deformed lids, or lids with gaps or other defects in the sealing gasket. Use new lids every time.

Photo 2.6 Self-sealing lid and screwband (top). Those closures are commonly used for water bath and pressure canners.

The lid goes on top of a jar and is held in place by the screwband that goes over it. The jar lids are usually *heated* in hot water and kept hot until ready to use. The screw band is secured finger tight as it does not play a major role in the sealing process.

- If rings are too loose, liquid may escape from jars during processing and seals may fail.
- If rings are too tight, air cannot vent during processing and food will discolor during storage. Over tightening also may cause lids to buckle and jars to break, especially with raw-packed, pressure-processed food.

When a jar is processed the sealing compound softens and covers the rim surface of the jar, yet it remains soft enough to allow the building air pressure to escape from the jar. As the jar cools, the compound hardens and the gasket forms an airtight seal. The slow hardening of the gasket is the reason why processed jars are left undisturbed for 12 hours to cool down. After cooling, the jars are inspected for a tight seal. The metal screwbands have served their purpose and may be left on the jars or removed and used in the next operation. *Do not tighten lids again after processing jars.* As jars cool, the contents in the jar contract, pulling the self-sealing lid firmly against the jar to form a high vacuum.

Screwbands are not needed on stored jars. They can be removed easily after jars are cooled. When removed, washed, dried, and stored in a dry area, screwbands may be used many times. If left on stored jars, they become difficult to remove, often rust, and may not work properly again.

Twist-On Jar

Lug is the closure system where the container has multiple threads and the lid has an equal number of lugs or tabs that grip the corresponding threads; also known as "twist-on". There may be three to six lugs depending on the diameter of the cap. Twist-on closures are mainly used by commercial producers, although they are available online. In their guides, the USDA mentions only glass with two part lids for home canning, however, all commercial products are canned in jars with a single lid only. Twist on jars (lug type) can be processed much faster by modern machines as the lid is closed with a part turn only. The lug closure (twist cap) can be easily removed and forms a good reseal for storage. Home canners in Europe use single lid closures and most of them are not familiar with American style two-part lid.

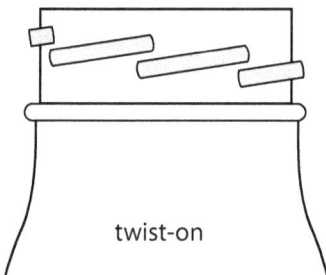

twist-on

Fig. 2.3 Twist-on jar.

Photo 2.7 Twist-on jar.

Sealing compound used in continuous thread and twist-on jars is known as plastisol and its composition may vary depending on the application.

Photo 2.8 Twist-on-lid. Twist-on lids are characterized by protruding lugs.

Be aware that standard one-piece lids are manufactured for hot-fill temperatures (around 180° F). Then, there are lids manufactured for hot water bath (212° F, 100° C) and the lids that can be used in pressure canners (over 212° F, 100° C). To summarize, do your homework first, then pick the lid suitable for your application.

Continuous thread single closure jars are commonly used in Europe for hot filling acidic foods such as jams or jellies. After the jar is filled, it is inverted a few minutes, then left to cool by itself. Inverting the jar allows plastisol to make contact with hot contents which softens the sealing compound. Upon cooling the vacuum forms inside and the jar is sealed.

Photo 2.9 Sealing jars filled with hot acidic food.

Fillmorecontainer, www.fillmorecontainer.com carries single piece canning lids with a vacuum indicating button for pressure canning. Those lids incorporate a sealing compound which has been designed for processing at higher temperatures.

Press and Twist Jar

Photo 2.10 Molded plastisol provides the seal along the top and side surfaces of the glass container finish.

Photo 2.11 Baby jar with 10 lugs.

Photo 2.12 Safety button is a standard feature on every baby jar.

Testing Sealed Jars

Most two piece lids will seal with a "pop"sound when cooling. When the canning jars have cooled down, test the lid. A canning jar is properly sealed if the lid is curved down or remains so when pressed. If it springs back and makes a clicking sound it is not properly sealed.

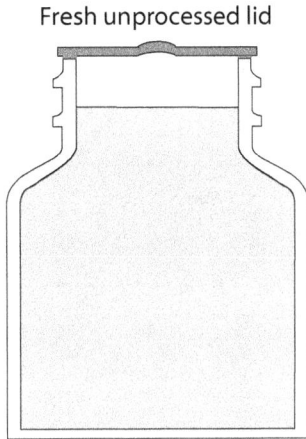

Fresh unprocessed lid   Heat processed lid

Fig. 2.4 Fresh unprocessed lid has a little curved up nipple in the middle.

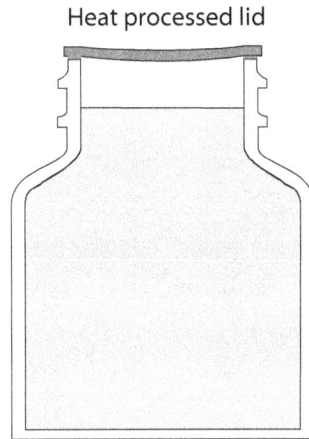

Fig. 2.5 Heat processed and properly sealed lid is curved down.

Another method is to tap the center of the lid with a spoon. A clear ringing sound means a good seal. A dull sound signifies that a jar does not have a tight seal or that the food is touching the lid. Hold the jar and look at it. If no food is touching the lid, the jar does not have a tight seal and cannot be stored. When the jars have completely cooled, the screwbands are removed. Due to humidity they may rust during storage making them difficult to remove. *The bands do not form the seal, the canning lids do.* Remove the screwbands, wash, dry and store them until needed again. Check the seals. Improperly sealed jars can be reprocessed within 24 hours using a new sterilized lid and the same processing time. If they are not remade, they should be refrigerated and used within 1 month.

Additional Equipment

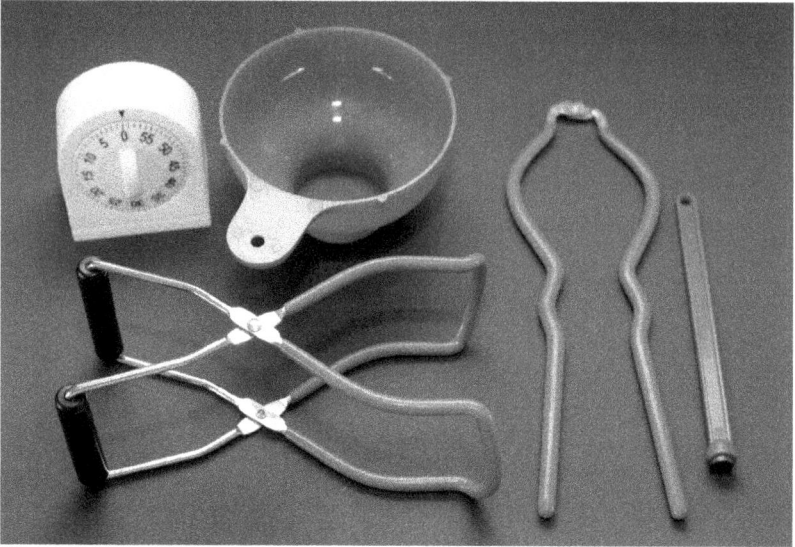

Photo 2.13 Additional equipment: timer, wide mouth funnel, jar holder, screwband remover (for stubborn bands), magnetic lid picker. Also needed is a scale, plastic knife, clean cloth, labels.

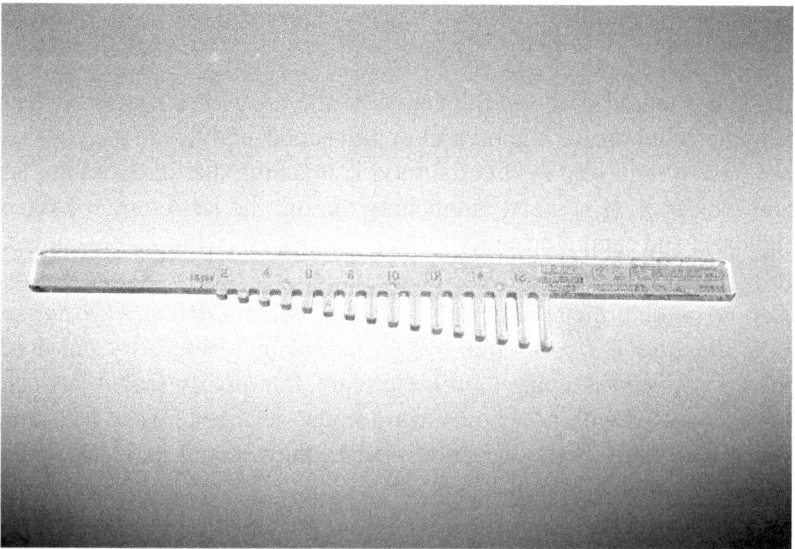

Photo 2.14 Headspace gauge

Metal Cans

The number of cans on the market is overwhelming and choosing the right can may be an intimidating task for a newcomer. Metal cans come in different shapes and sizes. The can sizes used in industry in the US are derived from nominal outside dimensions. While such dimensions may be expressed in inches, the custom is to use a conventional method in which three-digit numbers are used to express each dimension. The first digit indicates the number of whole inches in a dimension, and the second and third digits indicate the fractional inches as sixteenths of an inch. Some examples of can sizes can be found below.

211 x 400 means 2-11/16 x 4 inches
307 x 409 means 3-7/16 x 4-9/16 inches
404 x 414 means 4-4/16 x 4-14/16 inches

In oval cans, outside dimensions are used, the dimensions of the opening is stated first and followed by the height. The inches and sixteenths of an inch system is the same as with round cans. A 402 x 304 x 300, means that the oval opening is 4-2/16 x 3-4/16 inches and the height is 3 inches.

Metric system is much simpler and the sizes are defined in millimeters, for example:

round cans - 73 x 55 mm, 99 x 63 mm, 99 x 119 mm
oval cans - 90 x 103 x 217 mm.
Capacity, of course, is defined in milliliters.

There are cans which are suitable for certain applications, such as beverage cans, beer cans, soup cans or meat and fish cans. Food cans are lined up with enamel coating inside to prevent any reactions between metal and food which may affect the flavor of the product.

The most popular can is the basic round can with a consistent diameter and parallel walls, forming a cylinder. The tapered cans are of more interest to those canning at home. The tapered can has tapered body walls so that the diameter of the two ends are different. The advantage of this design is the cans can be nested during shipping, resulting in smaller packages and lower costs. They can also be stacked easily in storage. A hobbyist must consider the can size he intends to use with the size of the can sealer that can seal it. Metal cans are expensive when bought by the dozen, however, the price drastically drops when ordering a few hundred.

Industry Can Name	Capacity (Water)	Can Size	Dia. in inches	Height in inches	Size in mm
202	3.60 oz	202 x 204	2.13	2.25	54.0 x 57.2
202	4.80 oz	202 x 214	2.13	2.88	54.1 x 73.2
6 Z	6.00 oz	202 x 308	2.13	3.5	54.2 x 89.0
4 oz Mushroom	7.15 oz	211 x 212	2.69	2.75	68.3 x 69.9
8 Z Short	7.90 oz	211 x 300	2.69	3.00	68.4 x 76.3
8 Z Tall	8.6 oz	211 x 304	2.69	3.25	68.4 x 82.6
1 (Picnic)	10.9 oz	211 x 400	2.69	4.00	68.4 x 101.7
211 Cylinder	12 oz	211 x 214	2.69	4.88	68.4 x 124.0
8 oz Mushroom	15 oz	300 x 400	3.00	4.00	76.2 x 101.6
300	15.2 oz	300 x 407	3.00	4.44	76.2 x 112.7
300 Cylinder	21.8 on	300 x 509	3.00	5.56	76.2 x 141.3
Pint, salmon, tapered	16 oz	301 x 408	3.06	4.50	77.8 x 114.3
1 Tall	16.7 oz	301 x 411	3.06	4.69	77.8 x 119.2
303	16.8 Oz	303 x 406	3.19	4.38	81.0 x 111.3
303 Cylinder	21.8 oz	303 x 509	3.19	5.56	81.0 x 141.3
1/2 lb Tuna	5 oz	307 x 113	3.44	1.81	87.4 x 46.0
1/2 Pint (1 lb.), tapered	7.75 oz	307 x 200.25	3.44	2.00	87.4 x 51.2
2 Vacuum	14.7 oz	307 x 306	3.44	3.38	87.4 x 85.9
2	20.5	307 x 409	3.44	4.56	87.4 x 115.9
Jumbo	25.8 oz	307 x 510	3.44	5.63	87.4 x 143.0
2 Cylinder	25.8	307 x 512	3.44	5.75	87.4 x 146.1
1.25	13 oz	401 x 206	4.06	2.38	103.2 x 60.5
2.5	29.8	401 x 411	4.06	4.69	103.2 x 119.2
3 Vacuum	26.6 oz	404 x 307	4.25	3.44	108.0 x 87.4
3	35 oz	404 x 414	4.25	4.88	108.0 x 124.0
Tall No. 3	51.70 oz	404 x 700	4.25	7.00	108.0 x 177.8
5	59.10 oz	502 x 510	5.13	5.63	130.3 x 143.0
5 Squat	68.15 oz	603 X 408	6.19	4.50	157.2 X 177.8
10 (Gallon)	109 oz	603 x 700	6.19	7.00	157.3 x 177.8
12 (Full gallon)	138 oz	603 x 812	6.19	8.75	157.3 x 222.3

The capacity of a can is given in ounces of water at 68° F, 20° C.
1 US fluid ounce = 29.57 ml.
1 inch = 25.4 mm

Metal can sizes used in industry in the US are derived from nominal outside dimensions. Measurements are made of the empty round can before seaming on the packers' end. Using the outside diameter of the No. 2 can for calculating cylinder volume will tell us that the can is able to hold 23.5 liquid ounces. The can will actually hold only 20.5 ounces of water. Using the inside diameter will provide the correct number.

There are can-shaped containers made from aluminum foil and plastic laminates, there are rectangular-shaped cans and retortable pouches. Cans with easy-to-open rings on the lid are very popular. They all, however, require specialized equipment which is not available to a hobbyist. For home canning it is difficult to beat a round cylindrical can.

Popular Metric Cans and Their American Equivalent									
mm	52	65	73	83	99	105	127	153	168	176
USA	202	211	300	307	401	404	502	603	610	700

Metric size cans use the *inside diameter* to describe the can. For example No. 2 can, 307 x 409 converts to *87 (outside dia)* x 115 mm, however, it is classified as *83 (inside dia)* x 115 mm can in Europe.

Remember that every time you use a can with a different diameter, you need a different size chuck in your can sealer. When buying a sealer you have to specify the size of the can you will be sealing and the sealer comes with this size chuck and spacer pre-installed. Let us assume you have a 225 model sealer equipped for No. 2 can (307 diameter) and you decide that you want to can your salmon in smaller No. 1 cans (211 diameter). Well, you have two options: buy another 225 sealer equipped for No. 1 can, or buy an additional chuck for a No. 1 can. Needless to say, you have to switch over the chucks yourself.

Some machines like Ives-Way include different size chucks so you can seal a variety of different cans.

Types of Cans

Most cans are produced from tinplate. Tinplate is a thin sheet of steel, about 1/128-1/64 inch (0.2-0.4 mm) which is electrolytically coated with a very thin layer of tin on both sides. In addition, the interior of the cans is coated with a synthetic compound to prevent chemical reaction of the tinplate with the food.

The Three-Piece Cans

Three-piece steel cans are composed of the body, the bottom end and the top end (lid). The body is made of a sheet of tinplate, the sheet is made into a cylindrical shape and the ends slightly overlap. Then the ends are soldered and the soldering area is covered inside with a strip of coating for protection.

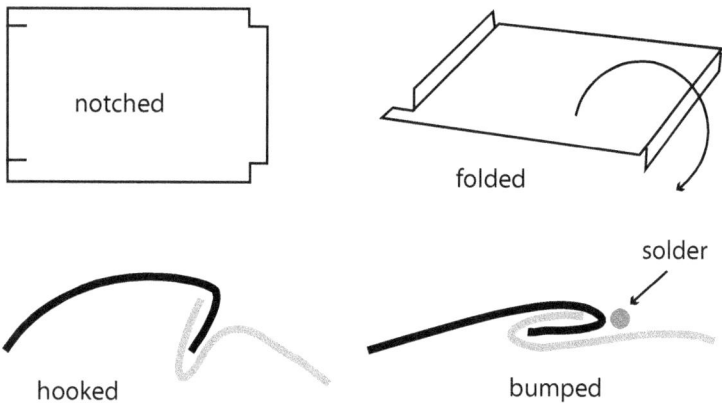

notched

folded

solder

hooked

bumped

Fig. 2.6 Soldered cans were made with lead solder. This procedure is not allowed anymore and lead solder cans cannot be legally imported into the US.

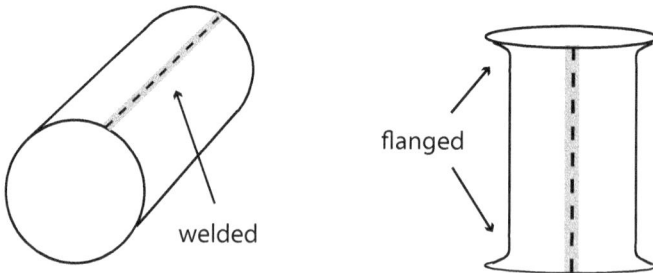

welded

flanged

Fig. 2.7 Currently, all 3-part metal cans are welded.

Photo 2.15 Protection coating strip over the weld inside of the can.

Photo 2.16 Right-the weld line outside of the can. The can is ribbed to increase its strength.

Photo 2.17 No. 2 can. Left: underside lid, center: the body, right: top lid. Enamel coating inside the can and underside the lid. Sealing compound visible on underside of the lid.

Photo 2.18 Underside view of the lid. Compound sealant in the perimeter area under the curl. The sealant is delicate, so the lids should not be washed in hot water.

After the body of the can is formed, one end is applied by the manufacturer, called the manufacturer's end, and the other called the canner's end, is applied by the canner using can sealer.

The Two-Piece Cans

The bottom end and the body are formed from one sheet of metal by stamping or a combination of stamping and deep drawing. It is easy to recognize a two-piece can because it has no seam at the bottom.

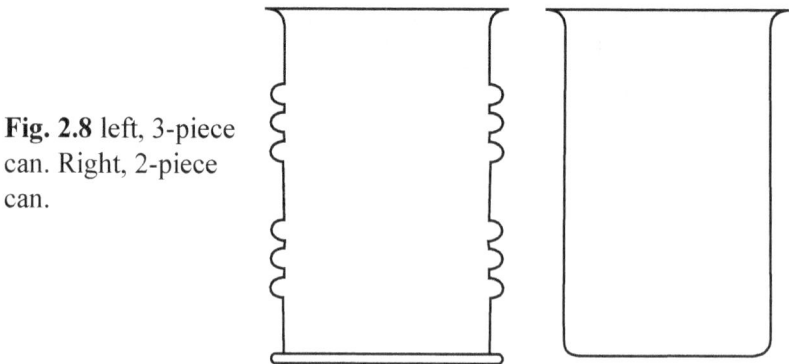

Fig. 2.8 left, 3-piece can. Right, 2-piece can.

There is a limitation to how much steel can be drawn and usually the side walls are twice the diameter of the can (1:2), for example, a popular ½ lb. tuna fish can is a two-piece drawn can of 87 x 45 mm.

Photo 2.19 Two-piece tuna has no seam on the bottom. The can has been flipped over (bottom on top). In both types of cans, the three-piece and the two-piece cans, the top end (lid) is applied with a can sealer.

Aluminum cans offer better deep-drawing capability so the walls of two-piece cans can be longer. They are more expensive, however, they offer benefits of low weight, an excellent heat transfer conductivity and strong resistance to corrosion.

Easy to Open Cans

Easy to open cans are extremely popular, however, they require specialized sealing equipment which is not available to a home owner.

Photo 2.20 Easy to open cans. Round cans made of steel, flat rectangular cans made of aluminum.

Fish cans are usually made from aluminum and they come in many shapes and sizes. Commercial producers need to create their own processing times, something at home canner is not able to do.

Tapered Cans

Two part tapered cans are the real space savers as they can be stacked inside one another. This lowers costs of shipping and facilitates storage. There are two sizes of tapered cans which are very popular for canning salmon: 307 x 200.25, 8 oz, and 301 x 408, 16 oz can. The fact that there is only one seam (the lid) to worry about makes them especially attractive for home applications.

Photo 2.21 Three different diameter cans. From left to right: 307 x 200, 8 oz tapered can, 301 x 408, 16 oz, tapered can, No. 2, 307 x 409, 20 oz, straight can. Straight cans cannot be stacked.

Glass Jar or a Metal Can, What's Better?

Commercial producers use metal cans not because they are pretty, they use them because they hold many advantages over glass jars. They can be processed faster, they don't break so easily and they are easier to cool resulting in a higher quality of the product.

Glass Jar	Metal Can
USDA Guides offer some information and recipes. The same information can be found in other publications and on the Internet.	*Almost no written information for a hobbyist.* A few expensive technical textbooks for people employed in the canning industry. Little information on the Internet.
Glass jars available in every large supermarket and online. Jars are standardized, almost all home canning is done in 2-part lid mason jars. The closure is not popular in Europe where most canning is done with a twist-on lid.	Metal cans are available only online. There are about 25 commonly used cans and choosing one may be a confusing task.
Canning in glass requires little knowledge and is easy to do. A perfect seal is assured and is easily verified by visual inspection and simple testing.	More knowledge is required by the operator. Some basic mechanical skills are needed to change over chucks and perform simple adjustments.
Contents are visible.	Contents are not visible.
Reusable, only lids need to be replaced.	Used only once.
No specialized equipment is needed as the canning operation is performed with a pressure canner, glass jars and common kitchen tools.	A can sealer machine is required. A different chuck must be installed to accommodate a can of different diameter. A spacer must be installed when using a can of a different length. Some adjusting is required.
Inexpensive to purchase in low quantity.	Expensive when buying a few cans. Cheap in large quantities.
Safer method as there are less chances for error.	Safe method, providing at home canners are willing to learn operating of a can sealer.
Due to the shape of the jar, solid shaped products will break upon removal.	Solid blocks such as luncheon meat or meat jelly can be removed intact.

Glass Jar	Metal Can
Glass breaks easy.	Strong. Packing machine can process thousands of cans per minute, something that is impractical using jars.
Glass breaks when exposed to a rapid change of temperature; after thermal treatment the jars are left to cool by themselves. Should not be handled during cooling (12 hours).	Can be cooled after thermal treatment. Can be handled at any time.
Seal needs time to harden.	Seal is set after sealing.
An additional cooking takes place during initial cooling of the jars. Overcooking results in a poor texture of the food.	Rapid cooling after thermal treatment stops cooking.
Jars should be kept in dark areas otherwise there will be change in food color and the fats will go rancid.	May be exposed to light.
When cooling temperature drops to 122-150° F, 50-66° C range, *thermophillic* spoilage bacteria which survived the thermal treatment can start growing again.	Can be rapidly cooled to 95° F, 35° C eliminating bacterial growth and additional cooking.
Commercial producers tend to process foods as fast as possible and at lowest possible temperatures. Glass jars are not suited for that.	Hundreds of cans can be sealed every minute. After thermal process cooling is accomplished without a delay in the same retort.
Glass jars are heavy. Glass does not corrode, but the screwbands do.	Cans are light. The outside surface will corrode in humid conditions.

The Verdict

- Canning in glass jars is easier to learn and jars are common. Jars cannot be rapidly cooled in home production.

- Canning in metal cans is a relatively simple method if one uses only one diameter cans. It requires basic skills to change chucks in order to process cans of different diameter and height. Adjusting rollers is also necessary as well as the knowledge of basic seam defects.

Chapter 3

Pressure Canners and Can Sealers

Low-acid foods must be processed in a pressure canner to be free of botulism risks.

The food may spoil if you fail to select the proper process times for specific altitudes, fail to exhaust canners properly, process at lower pressure than specified, process for fewer minutes than specified, or cool the canner with water. Small pressure canners hold four-quart jars; some large pressure canners hold 18 pint jars in two layers, but hold only seven quart jars individually. Pressure saucepans with smaller volume capacities are not recommended for use in canning. Small capacity pressure canners are treated in a similar manner as standard larger canners, and should be vented using the venting procedures.

Modern pressure canners are lightweight, thin walled kettles; most have turn-on lids. They have a jar rack, gasket, dial or weighted gauge, an automatic vent/cover lock, a vent port (steam vent) to be closed with a counterweight or weighted gauge, and a safety fuse. Pressure does not destroy microorganisms, but high temperatures applied for an adequate period of time do. A pressure canner is separated from the outside air by a forcefully shut lid.

Photo 3.1 Pressure canner, the lid is secured tight. In this All-American® 921 model, the six clamps hold the cover air-tight.

The success of destroying all microorganisms capable of growing in canned food is based on the temperature obtained in pure steam, free of air, at sea level. At sea level a canner set to a gauge pressure of 10½ lbs provides an internal temperature of 240° F, 116° C.

It is the amount of heat applied to the unit that regulates pressure.

If heat is continuously applied and the steam is prevented from escaping, the increasing pressure will reach the point when the vessel will explode. A pressure canner employs an adjustable pressure regulator, and of course, a safety valve (overpressure plug) which will open when the pressure reaches a dangerous level.

As the pressure is changing within the unit, the pressure regulator weight jiggles and sputters which is simply a process of constantly building up and releasing pressure to maintain the setting on the pressure regulator weight. Allow the canner to cool at room temperature until it is completely depressurized.

Pressure canners use two types of pressure gauges:

• Weighted gauge.

• Dial gauge.

It should be noted that a dial gauge needs to be checked for accuracy every year. Some canners come with a rubber sealing gasket and others make a tight fit without employing gaskets at all. All pressure canners come with detailed instructions for their use.

Weighted gauge has three settings: 5 PSI, 10 PSI and 15 PSI.

Pounds of Pressure PSI	Temperature	
	°C	°F
1	100	212
5	108	227
10	116	240
15	121	250

Two layers of jars can be processed at once, as long as the pressure canner is tall enough. Place a small wire rack between the layers to facilitate the circulation of steam. Make sure that there is 2 to 3 inches of water in the bottom.

Presto® Pressure Canners

National Presto Industries has been making Presto Pressure Canners for over 50 years. Presto Pressure Canners are made of high quality aluminum for great heat distribution. Don't confuse the pressure canner with smaller pressure cookers, also made by Presto®.

Photo 3.2 Pressure canner.

Photo 3.3 Dial pressure gauge.

Presto® pressure canners operate on a dial gauge principle. The dial gauge does not regulate the amount of pressure, it is just an indicator of the pressure inside. *The pressure is controlled by increasing or decreasing heat supply.* What is called the pressure regulator is actually a safety device to prevent pressure in excess of 15 pounds to build in the canner. The regulator sits loosely on top of the vent pipe, to the right of the dial gauge in the above photo. The real safety valve known as the overpressure plug is made of black rubber and can be seen on the left behind the dial gauge. Once the pressure increases to 15 PSI the pressure regulator starts to rock releasing pressure and maintain the setting at 15 PSI. If for any reasons the vent pipe becomes clogged, the regulator will not sense it so the pressure will increase well over 15 PSI. The overpressure plug, however, will pop out releasing the pressure. Check the vent pipe every time after placing the cover.

Photo 3.4 Air vent in down position, there is no pressure inside the canner.

There is a clever device called an air vent which is basically a cover lock and pressure indicator. When the pressure starts to build up, the air vent pops out and engages with the inside bracket of the canner. This prevents the cover from being opened when there is pressure in the canner.

Photo 3.5 Pressure builds up, air vent up, the canner is being vented, 0 PSI.

The heat is on, the air from the canner escapes through the vent pipe, which is not covered with a pressure regulator yet. Once the air exhausts out, the pressure regulator will be placed on top of the vent pipe where it will sit loosely. There is no indication on the dial yet, but the rising pressure and escaping air have already pushed the air vent up.

Photo 3.6 Presto regulator.

After venting canner for 10 minutes, the pressure regulator is placed on top of the vent pipe and the pressure will start increasing. The heat is usually left fully on.

During the 10 minutes venting procedure the steam forces air out of the canner. Air trapped in pressure canners lowers the temperature and results in under processing.

Once the pressure is about 2 PSI less than the desired setting, the heat is lowered and the pressure is raised in a slower, more controllable manner.

Photo 3.7 Low acid foods are canned at 11 PSI in a dial gauge pressure canners.

Photo 3.8 When the pressure reaches 15 PSI, the regulator starts to rock, maintaining the pressure at this setting. The regulator will rock only at 15 pounds of pressure..

Photo 3.9 Depressurizing canner. The regulator is off the vent pipe.

When the end of the processing time is reached, the heat is turned off. The canner should be removed from the electrical burner to prevent unnecessary heating, however, lift it up carefully, or you may scratch the surface of the glass top stove. If you are using a gas burner, just turn off the heat. The canner must cool by itself until the dial reads 0 psi and the air vent drops down. Now the pressure regulator can be taken off the vent pipe but let the canner cool for 10 more minutes before removing the cover.

Caution: escaping steam can scald you and the pressure control may be hot. Protect your hands with oven gloves and tilt the cover so steam escapes away from you. Using a jar lifter remove the jars one at a time. Place the hot jars on dry towels or a cooling rack. Let them sit undisturbed for 12 hours. Remove the screwbands and test the jars.

Regardless of the pressure canner used, processing time always begins when the pressure gauge registers the correct pressure. If pressure drops below the desired setting, it is necessary to bring the pressure to the correct setting and begin processing countdown from the beginning for the full amount of time.

All American® Pressure Canners

Established in 1909, Wisconsin Aluminum Foundry makes top quality pressure canners and can sealers. It is a heavy duty canner, without a rubber seal, operating on the weighted gauge principle.

Photo 3.11 Weighted gauge pressure regulator.

When the pressure canner reaches the desired pressure, the regulator weight will jiggle and sputter. The heat should be adjusted so that the regulator's weight will jiggle only about one to four times a minute. This indicates that the correct pressure is maintained.

Photo 3.10 All-American® 921 pressure canner.

The pressure regulator weight never requires adjustment or testing for accuracy. The pressure regulator weight can be selected to three pressure settings: 5, 10 and 15 lbs.

Photo 3.12 Weighted gauge on a vent pipe.

Photo 3.13 Dial pressure indicator.

Mirro Pressure Canners

Mirro pressure canners work on a weighted gauge principle. There is no dial pressure indicator. Use canners that are equipped with weighted gauge (5, 10, 15 lbs) only. Do not try to can in a single-control model. The weighted gauge should jiggle about 2-3 times per minute.

All pressure canners come with a rack. Placing jars directly on the bottom of the canner will likely crack them. Keep the jar upright at all times. Tilting the jar could cause food to spill into the sealing gasket of the lid and produce an imperfect seal. The jars may be stacked in the canner on two levels. Place another rack on top of the first row, then add more jars on top *or* offset the jars by placing one across two others.

American PSI	European bar	European kPa
10	0.6890	68.94
11	0.7579	75.84
12	0.8268	82.73
13	0.8957	89.63
14	0.9646	96.52
15	1.0335	103.42

Storage of the Canner

Clean and dry your canner. Store it with crumpled newspapers or paper towels in the bottom and around the rack. Place the lid upside down on the canner to protect the gauge, vent and sealing ring. Storing cans at high temperatures may result in *flat sour* - a type of spoilage where the can ends remain flat but the food turns sour. It is caused by heat loving bacteria which grow at 150-160° F, 66-71° C. Inadequate cooling after canning or storage at high temperatures promotes such spoilage.

Commercial Retorts

A pressure canner is a small canner used at home. The industry uses the term "retort." Retorts can process a few hundred or many thousands cans in one operation. It has to comply with the government regulations, for example each retort must have a temperature indicating device (thermometer) and a temperature recording chart.

The main difference between a pressure canner and a retort is the cooling capability of a commercial vessel. At the end of a thermal process the compressed air is injected inside and the retort is flooded with cold water. The cold water immediately lowers the temperature of the product and the compressed air prevents cans from buckling up and the contents of glass jars from boiling over. The temperature of the glass jars must be lowered gradually to prevent the glass from breaking due to the thermal shock. This pressurizing/cooling process allows for a precise control of cooking time and results in a higher quality, longer shelf life of product that can be produced at home.

Commercial retorts can be classified into the following groups:

- Still Steam Retorts (vertical or horizontal).
- Still Steam Retorts with Overpressure.
- Hydrostatic Retorts.
- Continuous Rotary Retorts.
- Batch Agitating Retorts.

In addition there are *Asceptic Processing and Packaging Systems*. Those systems are used for processing liquid foods which are pumped throughout the system. The asceptic system consists of three distinctive areas, all kept in sterile conditions:

- Equipment is sterilized.
- Food is cooked to the commercial sterility stage and stored in a sterile tank.
- Food is packed into sterile containers in a sterile room.

All mentioned above retorts and asceptic systems are covered in the FDA sanctioned Better Process Control course, see Chapter 1.

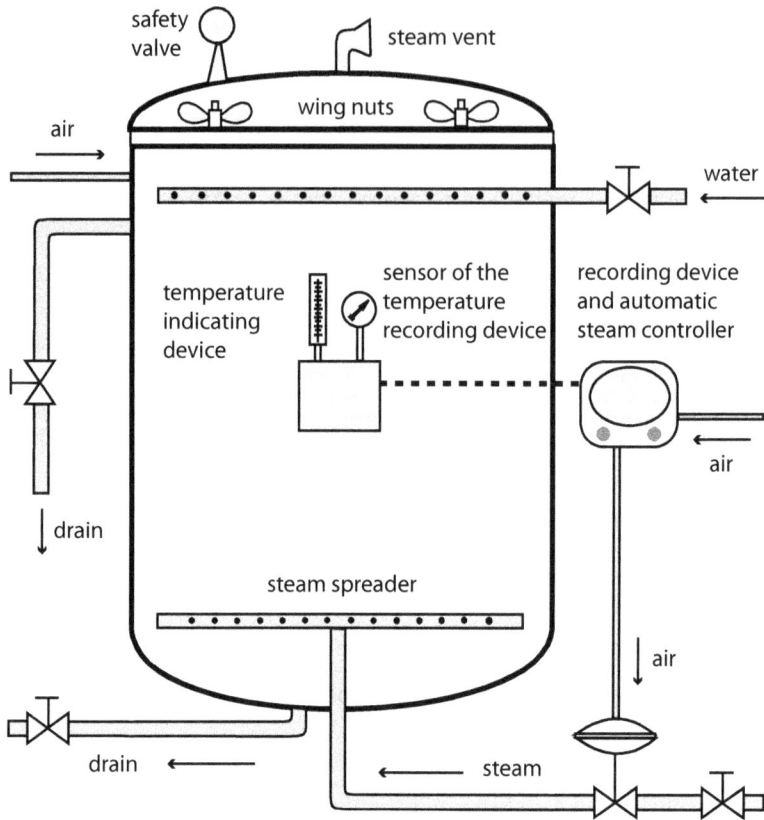

Fig. 3.1 Vertical still steam retort.

Like a home pressure canner, the vertical retort is loaded from the top. A wire basket holding hundreds of cans is lowered inside the retort.

Note: equipment and procedures for pressure processing in still steam retorts are covered in CFR 21, Part 113.40 *Equipment and Procedures*.

Can Sealers

The preservation of canned products requires hermetically sealed cans. The process of attaching the can end (lid) to the can body is called double seaming. This seam is formed by mechanically interlocking the outside end of the lid known as the "curl" with the top part of the can body known as the "flange."

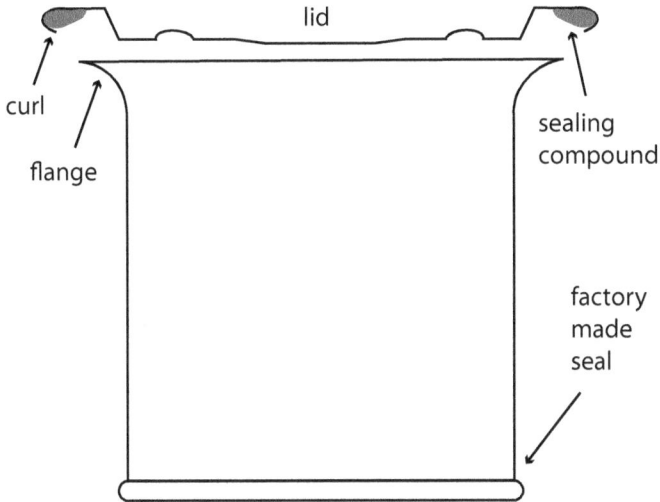

Fig. 3.2 Curl and flange form double seam.

Double Seam

A double seam is formed by joining the body of the can with the lid, which in trade is usually referred to as the cover or the end. The body flange and the curl of the lid interlock together and form a strong mechanical joint. This joint, although strong is not yet airtight and cannot be considered to be a hermetically formed closure. A sealing compound is attached to the curl of the lid. Compressing the lid and can body together forces the soft compound sealant to flow and fill any little spaces that might be present in a double seam and the combination of a double seam and the dealing compound makes a hermetically (airtight) formed closure. The amount of the compound sealer and its composition depends on the style of container and the method of sterilization.

Photo 3.14 There is a round flange on top of the body of the can. Can sealer will bend it together with a cover curl.

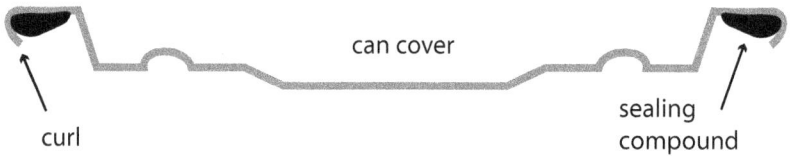

can cover

curl

sealing compound

Fig. 3.3 A cover carries a sealing compound inside of its curl.

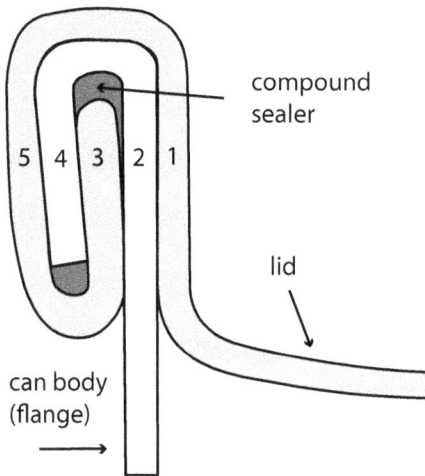

compound sealer

5 4 3 2 1

lid

can body (flange)

Fig. 3.4 Curl of the cover and the flange of the body form the double seam.

Each double seal consists of five layers of metal interlocked together: three from the lid and two from the body of the can.

Each sealer or "seamer machine" (industry name) has a:

- Base plate (turntable) - provides support for the body of the can.
- Seaming chuck - snugly holds the top of the can (lid) and acts as a support surface for the pressure of the rollers.
- A set of operation rollers - the first roller interlocks the lid with the body of the can. The second roller tightens and irons out the seam.

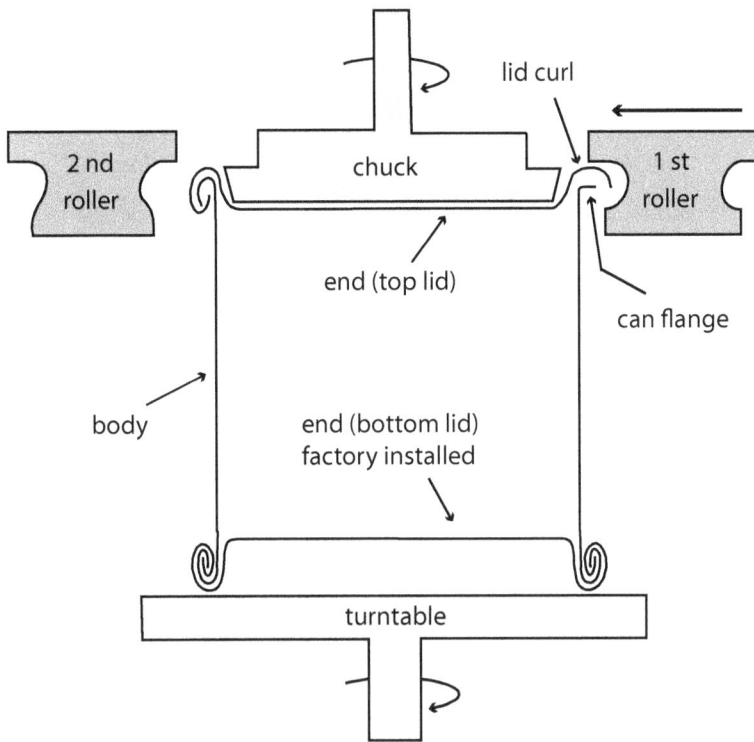

Fig. 3.5 Basic parts of a sealer.

Sealing is a critical operation in can processing. It must protect the contents of the can during thermal processing, and it must isolate canned food from micro-organisms and air during storage.

Photo 3.15 Ives-Way sealer. Chuck, 1st and 2nd roller and turntable. This turntable is free spinning and so are the rollers. The manual crank turns the chuck and the chuck turns the table (through the can).

Principles of Sealing Cans

There are two sealing machine designs:

- Can spin - the can is turning, the chuck is stationery.
- Can stand still - the chuck and the can are spinning.

Our discussion is limited to the "can still" design as the sealers designed for the home market are based on this principle.

Sealing cans involves three operations:

- Compression.
- 1st roller operation.
- 2nd roller operation.

Compression. Compression provides the necessary force for holding the can against the chuck and to feed the flange of the body into the first roller. Every closing machine has a base plate and a chuck. The snugly fitting chuck holds the lid in place and provides support for the seaming rolls pressure. The base plate is spring loaded and pushes the can up against the chuck.

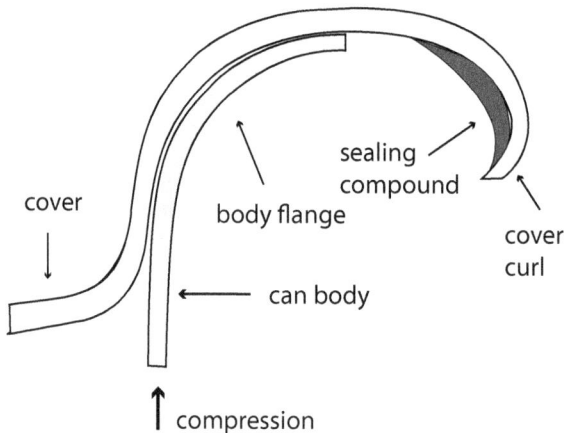

Fig. 3.6 Compression.

Chuck

All round cans are classified with diameter x height dimensions, for example 307 x 409 and 307 x 200. Every time the diameter (the first number) changes, a proper chuck must be installed. In the above example the cans have different diameters so 307 and 301 chucks must be used.

When a crank handle is turned, a set of gears transfers the movement to the chuck which starts turning. The serrations on the chuck force the lid to turn as well. The chuck fits snugly into the lid.

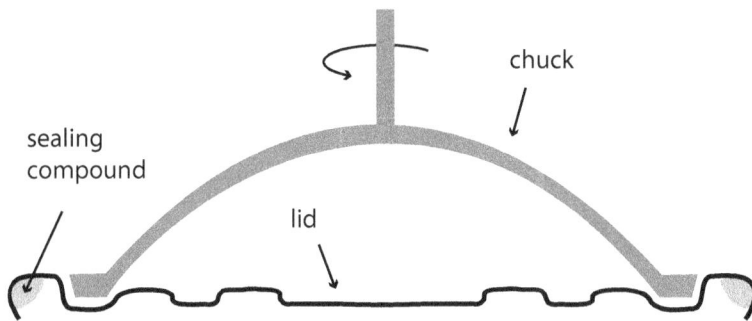

Fig. 3.7 Chuck spinning the lid.

Photo 3.16 Serrations help the chuck grab and hold the lid.

Photo 3.17 Chuck fits snugly inside the lid.

Turntable

Turntable, also known as base plate, is a free spinning disc that supports the can. The chuck propels the lid that sits on top of the body of the can. Without pressure the lid will continue to spin on top of the can and the can will remain stationery. The lid and the can must be compressed together and this is accomplished with a locking lever.

There are two designs for applying pressure:

- Ives sealer lever pushes the chuck downward towards the lid.
- All American sealer lever moves the turntable upward against the can.

The result is the same in both cases; there is pressure on a can from both ends and the whole system: the chuck, the lid, the can and the turntable starts spinning as one.

Photo 3.18 No compression, locking lever in upright position. Sealer ready for canning.

Photo 3.19 No compression, the can is loaded, but the chuck is still above the lid.

Can body and the cover are clamped together by engaging the locking lever.

Photo 3.20 Compression lever engaged. The can is compressed on both ends and ready for sealing.

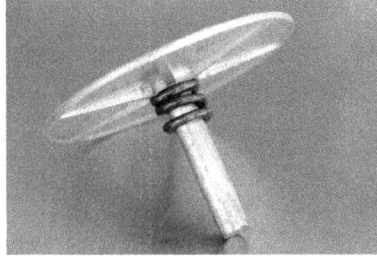

Photo 3.21 The spring applies pressure to the turntable.

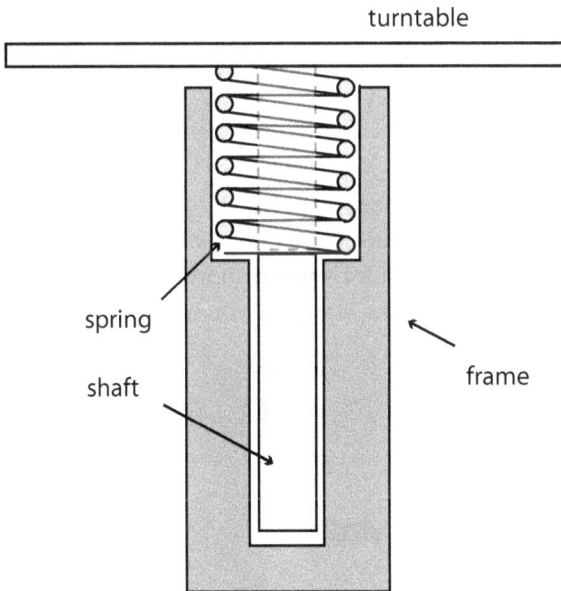

Fig. 3.8 Turntable rests on spring in sealer's frame.

The tallest can may get the correct amount of pressure from the spring alone, however, a shorter can of the same diameter will need a spacer. For example, a small salmon can with a 307 x 200.25 dimension and a No. 2 can with a 307 x 409 dimension. Both cans are sealed with 307 chuck, but they will need a different spacer to raise or lower the turntable.

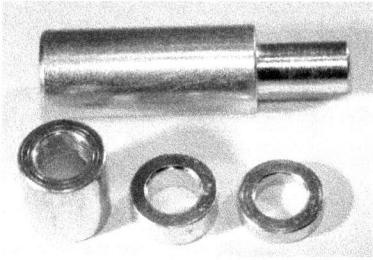

Photo 3.22 Spacers and extension table (top).

Photo 3.23 Spacer above the spring, Ives-Way sealer.

A few spacers may be placed together, however, there is a limit how much the shaft may be raised before it starts to wobble. The table extension spacer is needed for sealing short cans. When ordering cans with different heights, make sure that you order the correct spacers.

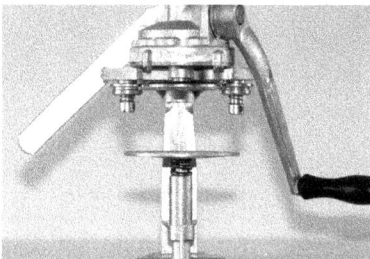

Photo 3.24 Turntable extension.

Photo 3.25 Turntable has grooves which confine the can in one area.

Roller Operation

The 1st and the 2nd roller have different specially countered grooves. During seaming only one roller makes contact with the can. When the lid is placed on a can, the rollers are away. Then the 1st roller makes a pass, but the 2nd roller is waiting its turn. The 1st roller moves away and the 2nd roller flattens the seam.

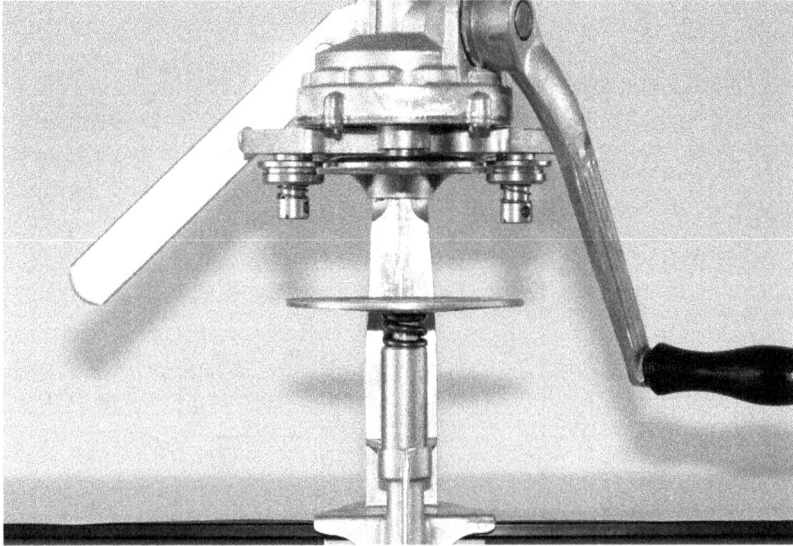

Photo 3.26 Seaming rollers, Ives-Way sealer.

1st Roller Operation

The curl of the lid is interlocked with the flange of the lid. The first operation should not be too loose or too tight, since there is no way to correct it later.

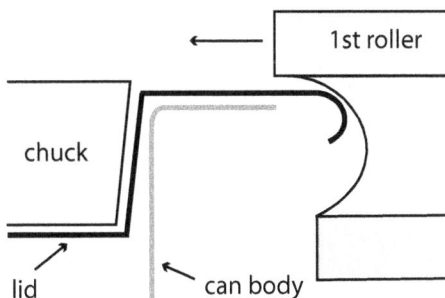

Photo 3.9 A curl of the lid occurs because a lid is wider than the can flange and the curl starts to turn around first.

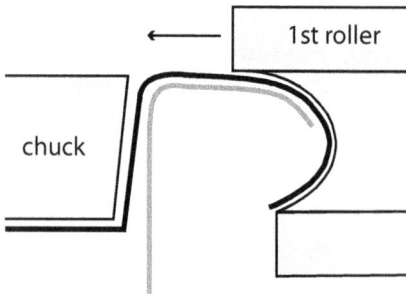

Fig. 3.10 The flange of the can follows as the roller moves towards the chuck.

The 2nd Roller Operation

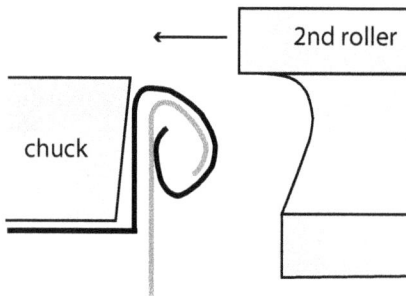

Fig. 3.12 Second roller begins to approach the hooks.

2nd operation roll seam

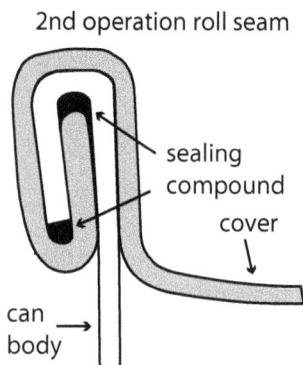

sealing compound

cover

can body

Fig. 3.14 Completed second operation.

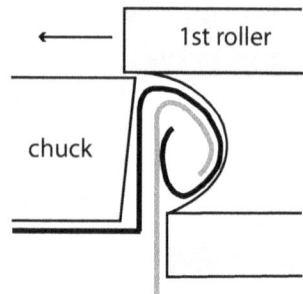

Fig. 3.11 The lid is interlocked with the can but the can is not hermetically sealed yet. The lid might turn around on top of the can if pressed hard enough.

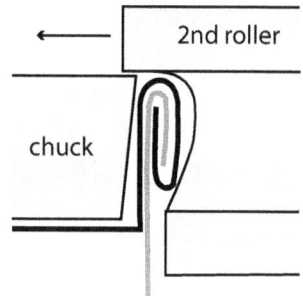

Fig. 3.13 Second roller compresses interlocked hooks.

The interlocked layers of cover and can body flange are compressed, and the sealing compound is squeezed into open spaces to complete the hermetic seal.

Ives-Way Sealer

Ives-Way is a wonderful little sealer that has been around since November 28, 1939. This is when Clifford E. Ives was granted a patent (2.181,237) for his machine. It has had enough time to be improved upon and be perfected. Since 1989 it has been manufactured and distributed by Ives-Way Products Inc,. located in Round Lake Beach, Illinois. The Ives-Way sealer can seal cans of 6 different diameters and adjusts from flat tuna fish cans to the taller fruit cans.

The sealer comes partially assembled with a set of different chucks and spacers. A detailed installation and operating manual is included. The company believes that it will benefit a new operator if he learns how to assemble and adjust the machine. The advantage of this approach is that by the time you assemble the sealer you will be well acquainted with its operation. You will know how to adjust the rollers, check clearances, change spacers and chucks, oil the crucial parts and always maintain it in top operating condition. Some distributors will assemble the sealer with a chuck of your choice for a small fee, however, we feel that it is best you assemble the sealer yourself.

Photo 3.27 Ives-Way 800 sealer comes with 301 and 307 chucks, spacers, set up gauges and the manual.

Photo 3.28 Ives-Way 800 sealer.

Different models come equipped with different chucks, however, the machine remains the same. What it means is that a person familiar with the unit can purchase additional chucks and will be able to can the six most popular can sizes. Many companies distribute this sealer, but information about the machine is the best kept secret on the internet; the reason being that until today the company has no visible presence on the internet and prefers to do business the old fashioned way. This works well as the company answers customer's telephone calls and will answer any questions about their range of products.

Food cans come in all shapes and sizes, so one particular sealer cannot seal all of them. Sealers that are designed for home canning will seal the round cans only. All sealers can accommodate optional chucks, and that will allow to seal cans of different diameters.

What follows is a brief description of the Ives-Way sealer. The sealer consists of the following parts: chuck, turntable, 1st and 2nd roller, locking lever, crank, associated hardware (pins, nuts, spacers), and adjustment wire gauges.

Photo 3.29 Chucks. By installing different chucks, the same Ives sealer can seal cans of six different diameters.

Keep in mind that every time a chuck is installed, both pressure rollers must be readjusted as well.

Make sure you know on which side of the chuck the 1st and the 2nd roller is located. Read the operating instructions. On the Ives sealer the 1st roller is to the *right* of the chuck. Ives-Way sealer uses a clever window indicator which displays each stage of the sealing process the can is experiencing.

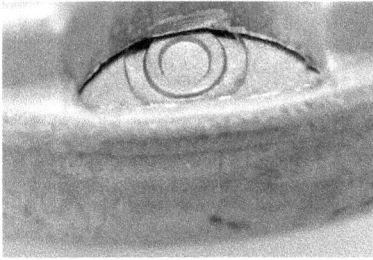

Photo 3.30 "0" - Ready for sealing. Operator starts cranking the handle.

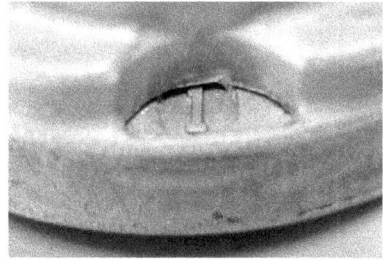

Photo 3.31 "1" - engaging 1st roller.

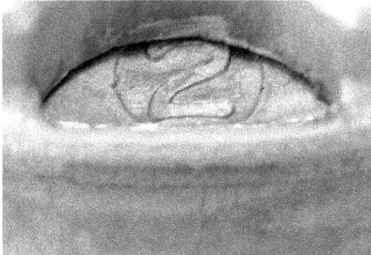

Photo 3.32 "2"- engaging 2nd roller. The process ends when "0" shows up again.

All American Can Sealers

Established in 1909, Wisconsin Aluminum Foundry makes top quality pressure canners and can sealers under the All American® label. All American® Can Sealers include two types of sealers:

- All-American Master Can Sealers to seal your choice of either No. 2 (307 x 409) or No. 3 (404 x 414) cans. Additional chucks are available for sealing other sizes, at an extra cost. The largest can size for All-American Master Can Sealers is No. 3 (quart size).

- All-American Senior Can Sealers are set up to seal No. 10 & No. 12 cans, and can be adapted for smaller sizes. Those models also come in an electric version which are capable of sealing 150 cans per hour.

All American Can Sealers are *heavy duty sealers* designed for use by small commercial canners, custom canners, school canneries, lunch rooms, experimental laboratories, gift packaging centers, and home canners. No skill or experience is required to perfectly seal tin cans automatically. It is unlikely that a hobbyist will need a bigger size than No. 2 can for sealing low acid products, and the 225 Master Can Sealer is well suited for the job. Larger models are more expensive and would be used by Commercial Canners, Custom Canners, School Canneries, Lunch Rooms or Experimental Laboratories. In addition, the recipes that were published by the USDA for home canning were tested for No. 2, No. 2.5 and No. 3 cans.

Make note that 225 sealer will not seal smaller than No. 2 cans.

Photo 3.33 Compression off. All-American Master 225 sealer. The lever in off position, compression off.

Photo 3.34 Compression on. All-American Master 225 sealer. The lever in on position, the table is raised, compression on.

Evaluation of the Double Seam

The shape and conformation of the finished seam are determined by the design of the seaming rolls and the taper of the chuck. Roll shapes may be changed to accommodate different thicknesses of the can material. The roll shapes, the pressure adjustments of the rolls and the base plate will determine the shape, the dimensions and the integrity of the double seam.

At first sight, the shape of the double seam may appear to be satisfactory, however, one or more of the internal structures may be out of limits. As a result, a hermetic seal will not be produced. The commercial sealing machine may close 1000 cans each minute, so a defect, if not recognized soon, may result in thousands of defective closures. To prevent this from happening, a trained mechanic is continuously inspecting the production.

- **A Visual Inspection** of closures shall be performed at sufficient intervals which may discover more obvious defects. The interval between visual inspections cannot exceed 30 minutes of continuous sealing machine operation.

- In addition, USDA/FSIS regulations require **Double Seam Teardown** examinations which shall be performed at intervals of sufficient frequency, not to exceed 4 operating hours.

A sealer designed for home canning operates at lesser capacity, however, an understanding of the double seam inspection procedures will make a hobbyist more knowledgeable. The following drawing depicts crucial dimensions of a completed double seam. They shall be measured with a quality micrometer capable of taking measurements with an accuracy of 1/1000 of inch.

The required seam measurements: cover hook length, body hook length, width (length, height), tightness (observation for wrinkle), thickness.

The optional seam measurements: overlap (by calculation), countersink.

The detailed requirements are listed in 21 CFR, Part 113.60.
The question may arise as where to obtain the technical dimension of a properly formed double seam. The can manufacturer has all technical data and his distributors have access to this information.

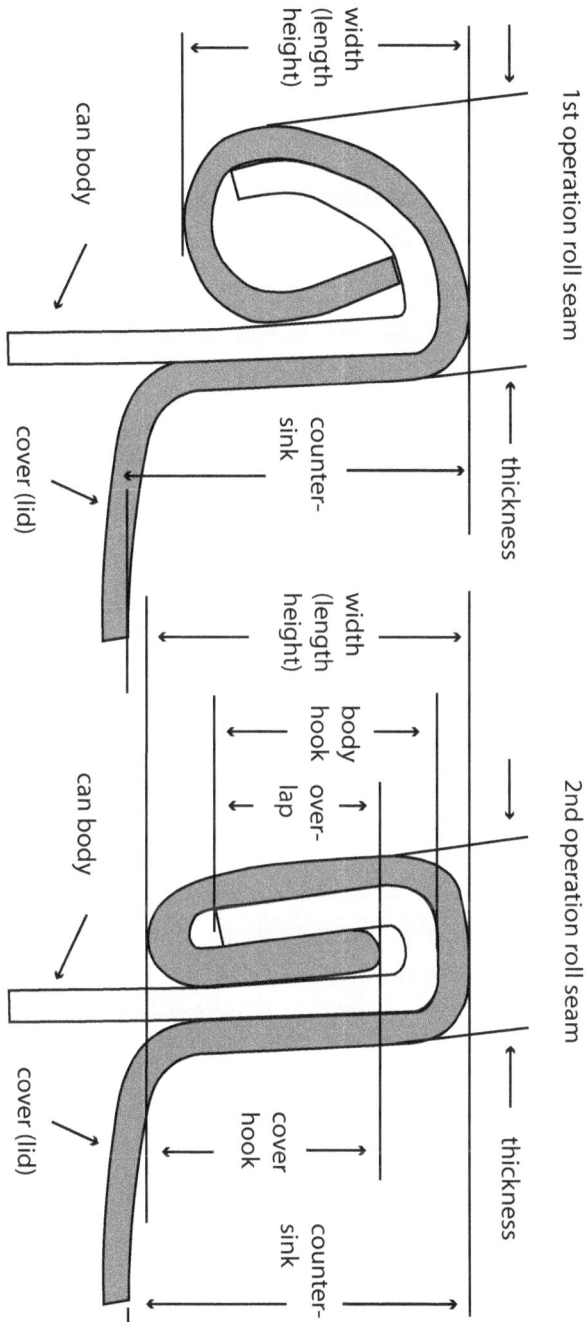

Fig. 3.15 Seam measurements.

Government regulations allow for two different types of the double seam examinations:

1. The micrometer method. The invention of the micrometer is attributed to Jean Laurent Palmer of Paris in 1848 and the micrometer caliper which was introduced to the mass market by Brown & Sharpe in 1867, bringing the instrument into local shops.

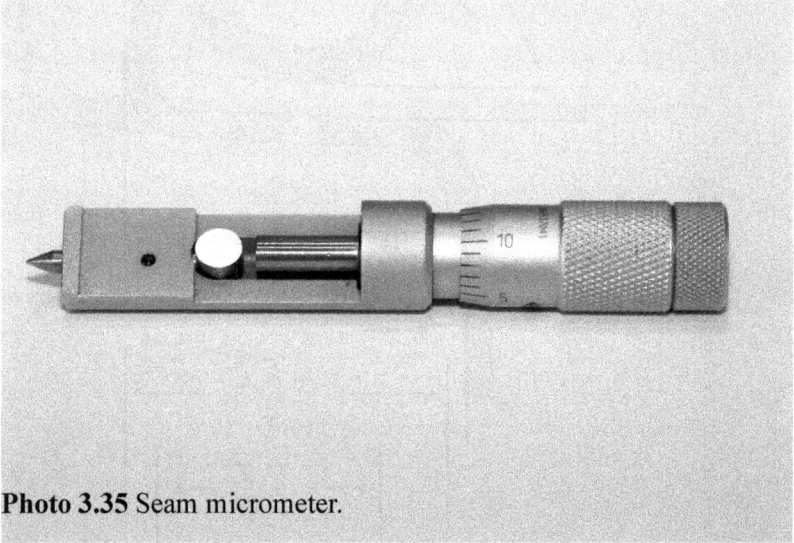

Photo 3.35 Seam micrometer.

The seam measurements are taken in thousandths of an inch so precision tools are necessary. A standard micrometer with a round anvil may be used, however, the seam micrometer has a stem that has been specifically designed for taking seam measurements.

Closing machines produce strong double seams when properly set up and adjusted. Because there are different manufacturers of sealing machines and metal cans, it is impossible to have one set of specifications that would apply to all machines and all sizes of cans. For that reason, commercial plants obtain specifications and technical support from the representatives of the companies that manufacture the closing machines and the cans. The following seam data is offered in good faith as typical reference and not as specification or requirements: cover and body hooks each are about the same length, and they fall into the range of from 0.076 to 0.084 inches in length. These tolerance ranges are for the cans from 301 to 404 diameters. Countersink depth is around 0.125 inch.

2. The optical method. The seam projector is a computerized accurate machine that not only displays an ex-ray-type image on the monitor, but is also capable of precisely calculating the seam measurements. Due to its cost, it is unlikely that the seam projector will be used for home canning.

Commercial plants use computerized seam projectors. Quality By Vision is a global leader in the development and manufacturing of Quality and Process Control systems for the canning industry and its suppliers. Their SEAMetal HD double seam vision system provides an unprecedented look at the double seam. A sharp and clear image is supported by very precise measurements.

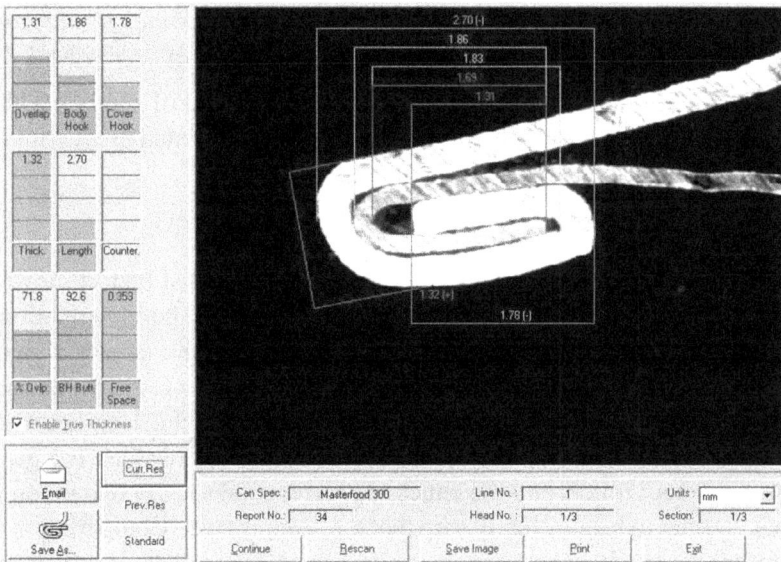

Photo 3.36 SEAMetal HD double seam computerized seam projector system.   *Photo courtesy: Quality By Vision*
www.qbyv.com

The system runs under Windows operating systems (including Vista) and connects to the computer using USB. The software measures: body thickness, cover thickness, seam thickness, seam length (seam height), body hook, cover hook, wrinkle (tightness), seam gap, countersink, body hook butting, cover hook butting, overlap, overlap %, (countersink - length), vee, and many more.

Quoted from the "Canadian Food Inspection Agency, Metal Can Defects Manual - Identification and Classification":

"It is extremely important to recognize and understand that the quality of the double seams cannot always be judged on measured dimensions alone. Visual inspection for tightness and visible abnormalities are equally important. Dimensions outside the can maker's guidelines do not necessarily mean that seam integrity is compromised. It means that the seam should be carefully evaluated. Final judgement must be based on the amount of deviation along with all of the other measurements and observations.

Can makers supply guidelines to their customers and indicate frequency of tests as well as points of measurement. These can makers' guidelines recognize the need to check certain attributes at certain points on the can. Not all tests need to be performed at every check."

For those who want to study the subject deeper a list of sources with additional information about can seam defects is provided in the Links of Interest section.

Teardown Examinations

As mentioned earlier, in a commercial plant cans must be torn-down and examined at intervals not exceeding 4 hours. The examination should be also performed after a shut down. A home canner is not bound by those regulations and he is not expected to tear up the can every time he uses a can sealer. If he changes the chuck to accommodate a different diameter can, however, he must readjust the gap between both rollers and the chuck. This is a good time to tear up one can in order to be absolutely sure that the seam is tight. Comparing the closure made with a home sealer to the bottom seam made in a factory is not recommended as the two will differ. The commercial seaming heads work differently and often employ more than one set of the 1st and the 2nd rollers.

In commercial operations a specially designed power-driven seam saw cuts cans fast and clean so they can be examined. A home canner will use commonly available tools such as a hack saw, file or handheld rotary tools.

Teardown Procedure

1. Remove the center panel of the cover.
2. Tear off the strip of the cover left by the can opener.
3. Cut the can body to expose the cross section of the seam.
4. Examine/measure the seam.

Tools

In order to measure the seam accurately the can must be torn down.

Note: measurements like countersink depth, seam width and thickness must be taken *before* the can is cut.

Can Opener. A conventional household can opener should not be used as it will distort the seam and the measurements will be off. A special can seam opener will cut the center part of the lid without damaging the seam.

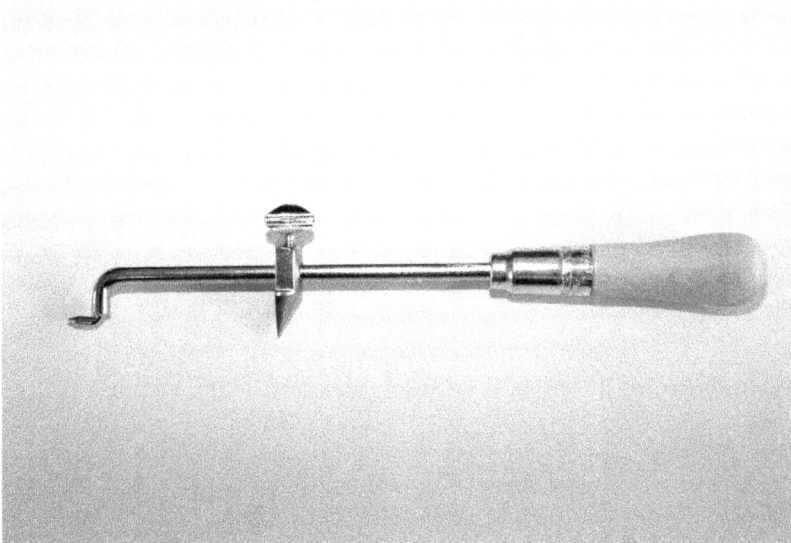

Photo 3.37 Can seam opener.

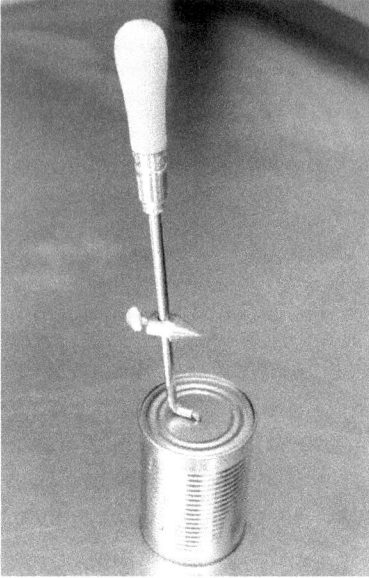

Photo 3.38 Insert can opener.

Photo 3.39 Adjust diameter.

Photo 3.40 Cut.

Photo 3.41 Lift the cover.

The center part of the lid can be cut out with other tools as long as the seam is not damaged. Punch a hole in the lid and then cut the lid out using aviation snips.

Photo 3.42 Oversize bottle cap opener.

Photo 3.43 Punching hole.

Photo 3.44 Aviation snips.

Fig. 3.45 Center part of the lid is removed.

You can tear off the remaining strip of the cover now with nippers or a suitable pliers. Stay close to the edge of the can, grab the cover and start twisting the pliers. The cover will start forming a little spiral. If it becomes too bulky, cut it off with aviation snips and continue as before.

Fig. 3.46 Stripping the cover.

Fig. 3.47 Stripped can. The lid has been removed without inflicting damage to the seam. The protruding little lip on the circumference of the can is the beginning of the cover hook.

Exposing the Seam

Visual inspection of the seam is a simple procedure. Cut the center part of the lid out as described earlier. You don't need to strip away the remaining part of the lid. A home canner can use of of the following methods:

- Hacksaw.
- File.
- Rotary tool with a cut off disc.

Hacksaw. Even if a fine blade is chosen (32 teeth per inch), cutting through the seam using a hacksaw is difficult unless the can is strongly supported. This will require a large vise or pipe holding equipment.

File. The filing method presents a little problem as the cut is not perfectly clean. The metal shavings fill the gaps in the seam and obstruct the view. You will need to obtain a lath file which has one smooth surface and will not interfere with the seam. Filing is easier as one hand can hold the can and another cuts with the file.

Cut through the seam, about 1½ inch away left from the vertical side weld. Then make another cut about 1 inch away from the first one.

Photo 3.48 Filed grooves.

Push from inside on the cut section and it will bend away from the can. The seam can be examined with a 10X jewelers loupe.

Photo 3.49 Seam section ready for visual inspection.

Examine the cross section of the seam with a rotary tool. A cutting disc leaves a clean cut. There is a variety of rotary tools and grinders that will do the job. The simplest solution is presented below: a common drill with an abrasive cut-off disc.

Caution: This is a power tool so pay attention to your safety. Goggles, gloves and securing the can are a must.

Photo 3.50 Abrasive cut off disc.

Photo 3.51 Using common drill equipped with a cut-off disc.

The home canner will probably not go much further beyond the visual inspection. To measure the body and cover hooks and to determine the tightness of the seam, the more detailed tear-down procedure is required. The body and cover hooks are compressed together but they need to be separated in order to take measurements. The complete stripping of the cover disengages the hook which is still attached to the body hook (see A).

Make two file cuts through the seam into the body of the can, about an inch apart. File through the first thickness of seam. Do not file into the body hook as this will shorten the body hook measurement. This removes the top part of the cover hook (see B). Tap the edge of the file on the side of the filed section to knock it down and remove it. This is the hardest part, but take your time in order not to damage the seam.

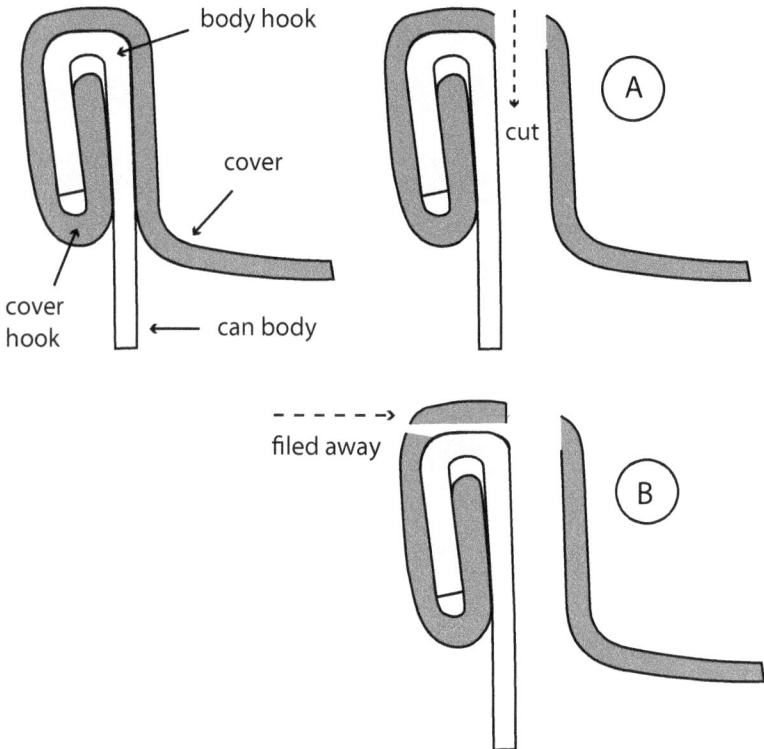

Fig. 3.16 Disengaging hooks.

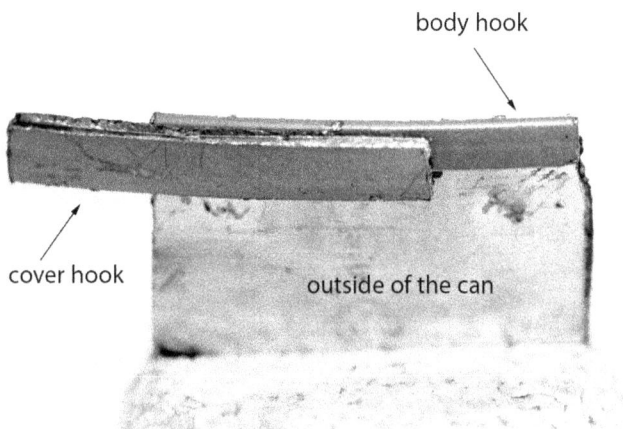

Photo 3.52 View at the Fig. 3.16 - circle B looking at the seam from the outside.

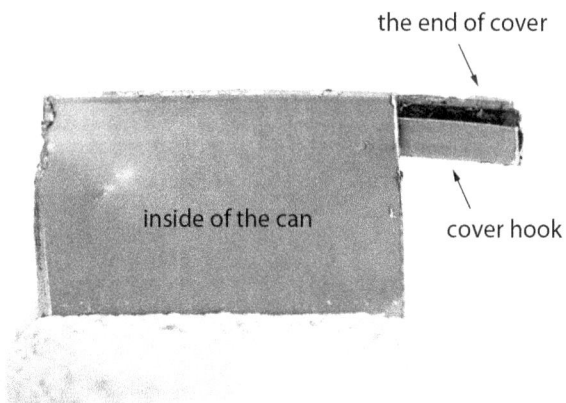

Photo 3.53 View at Fig. 3.16 - circle B looking at the seam from the inside of the can.

Photo 3.54 A part of the cover hook.

Fig. 3.55 A part of the cover hook still attached to the can. The can has not been filed yet.

What follows are the most common defects that may be observed with a naked eye during tear up inspections.

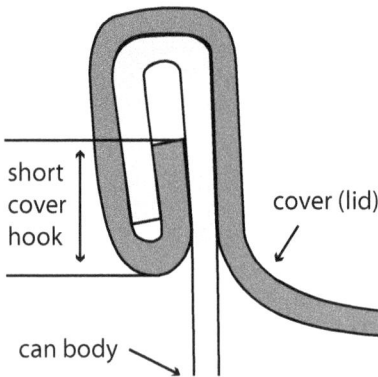

Fig. 3.17 Short cover hook.

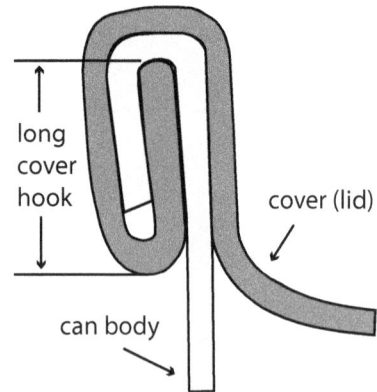

Fig. 3.18 Long cover hook.

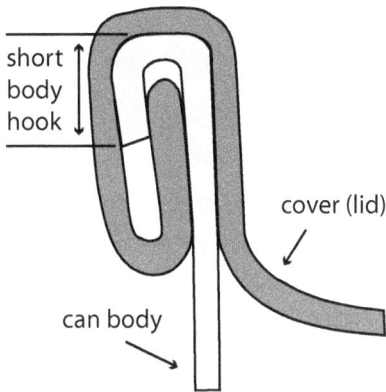

Fig. 3.19 Short body hook.

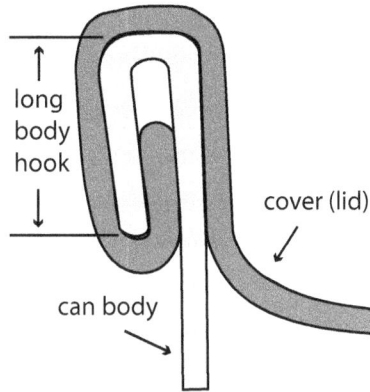

Fig. 20 Long body hook.

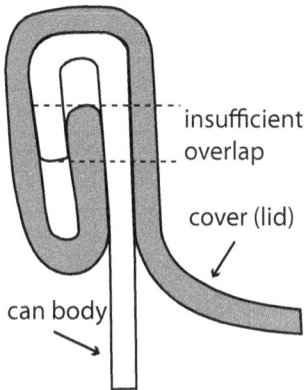

Fig. 3.21 Insufficient overlap.

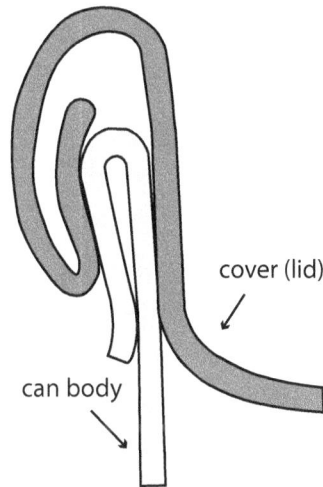

Fig 3.22 False seam.

Some defects are more important than others, for example a false seam or insufficient overlap may be considered critical defects. There are many more defects like sharp seams, bumped seam, droop, vee or knocked down flange.

Caution: If the seam looks in any way different than usual, throw the can away. In 3-piece cans you may compare your seam with the factory made bottom seam, however, keep in mind that it was made on a different machine. Can sealers should be checked before using to see that they are adjusted to make a good seal. One method of testing, popular amongst home canners, is to partly fill a can with water. Seal the cover, then immerse the can into hot water deep enough to cover the can. Use a jar lifter or another can to hold the tested can submerged. If air bubbles come up around the lid of the can, the seam is not tight and the sealer needs adjusting.

Caution: according to experts this method will not reveal microscopic holes. It is clear that this method will not be accepted by the FDA/USDA inspectors.

Note: much more information about seams and their defects can be found at the excellent site The Double Seam www.doubleseam.com

Useful Tools

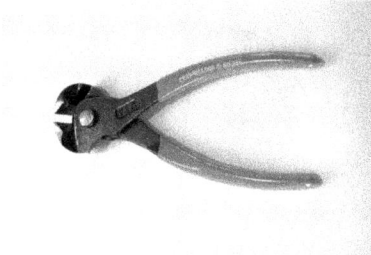

Photo 3.56 Nippers are used to tear off the can seams. No. 5 is a good choice.

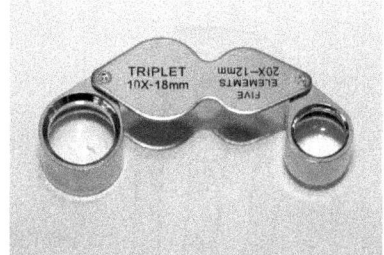

Photo 3.57 Jewelers loupe. 10X magnification is all that is needed.

A home canner is not going to invest thousands of dollars into a computerized seam projector. For a few dollars, however, he can buy a loupe which will enlarge the details of the seam.

Fig. 3.23 Canning vacuum gauge is used to check the vacuum created after a jar or can is sealed. Gauge needle is pushed through the cap, the rubber collar forms a temporary seal to allow for measurement. Measures vacuum from 0-30 inches of mercury (in Hg).

Chapter 4

Microbiology of Canned Foods

Home canning of jams is a very simple and well known process. It has been practiced for generations, it requires basic skills and equipment and it has been safely performed all over the world.

But home canning of meat, poultry, fish and vegetables is a more involved process that requires a sound understanding of microbiology and basic principles of physics. The fact that your grandmother processed meat and preserves in the same water boiling pot, does not mean she did it right! She did it *wrong* but she was lucky, and nobody in the family died. Many people had less luck and their obituary said: "died of natural causes" as that was the only explanation that the doctor could think of at the time. Having mentioned the dangerous consequences, we have to mention there are two completely different methods of canning foods and they both require different temperatures, times and equipment. The same glass jar or can may be used for both processes but the processes are entirely different.

- High-Acid Foods - jams, jellies, juices.
- Low-Acid Foods - meat, poultry, fish, vegetables.

The commercial production of low-acid foods is highly regulated by the Food and Drug Administration (FDA) and the United States Department of Agriculture, Food Safety and Inspection Service and rightly so, since people get sick and die from eating tainted food. Even with all the regulations we have a few massive recalls of processed foods every year. And if you check the headlines, the recalled foods were not jams, juices or preserves, they were either meats or vegetables.

The logical question arises as to what makes those low-acid foods so special? Well, the explanation is very simple, but it requires an understanding of a few basic concepts of microbiology. In other words, those tiny, invisible to the naked eye microorganisms decide how the food must be processed.

Safety of Canned Foods

Food begins to spoil soon after it is harvested or slaughtered. This spoilage is caused by microorganisms or by internal chemical changes which are caused by enzymes. Microorganisms can be destroyed and enzymes can be deactivated by heat treatment and this is the reason why thermal processing is the focal point in canning technology. Bacteria types vary and display different preferences for temperature at which they will grow or die. Understanding bacterial behavior is crucial for a better understanding of the heating and cooling process. Microorganisms such as molds, yeasts and bacteria spoil food, even at refrigerator temperatures. It is quite obvious that we have to do *something special* to canned food if it is to remain for 2-3 years without refrigeration. If not killed the microorganisms will find the moisture and nutrients inside of the canned food and they will multiply. Some bacteria will simply spoil the food, others might produce toxins. Clearly, we have to protect ourselves and either kill the bacteria or create such conditions that they will be unable to grow.

Food safety is nothing else but the control of bacteria. To control them effectively we have to first learn how bacteria behave. Let us make something clear, it is impossible to eliminate bacteria altogether. Life on the planet will come to a halt. They are everywhere; on the floor, on walls, in the air, and on our hands. All they need to grow is moisture, nutrients and warm temperature. They all share one thing in common: they want to live. Given the proper conditions they will start multiplying. They don't grow bigger, they just divide and divide and divide until there is nothing for them to eat, or until conditions become so unfavorable that they stop multiplying and die.

Microorganisms can be divided into three groups:

- Molds - are easy to kill and most will die below the temperature of boiling water; they are of little concern when canning meats.

- Yeasts - are easy to kill and most will die below the temperature of boiling water; they are of little concern when canning meats.

- Bacteria - these microorganism can be dangerous and must be properly dealt with.

It is commonly believed that the presence of bacteria creates immense danger to us but this belief is far from the truth. The fact is that a very small percentage of bacteria can place us in danger, and most of us with a healthy immune system are able to fight them off.

Bacteria Growth in Time

Under the correct conditions, bacteria reproduce rapidly and the populations can grow very large. *Temperature and time are the factors that affect bacterial growth the most.* Below 45° F bacteria will grow slowly and at temperatures above 140° F they start to die. In the so called "danger zone" between 40-140° F many bacteria grow very well. For instance, the infamous *E.coli* grows best at 37° C (98° F) and *Staph.aureus* at 30°-37° C (86°-98° F).

When bacteria grow, they divide and increase in numbers, not in size. Looking at the table it becomes clear what happens to a piece of meat left on the kitchen table on a beautiful and hot summer day.

# Bacteria	Elapsed time
10	0
20	20 minutes
40	40 minutes
80	1 hour
160	1 hour 20 min
320	1 hour 40 min
640	2 hours
1280	2 hours 20 min
2560	2 hours 40 min
5120	3 hours
10,240	3 hours 20 min
20,480	3 hours 40 min
40,960	4 hours
81,920	4 hours 20 min
163.840	4 hours 40 min
327,680	5 hours
655,360	5 hours 20 min
1,310,720	5 hours 40 min
2,621,440	6 hours

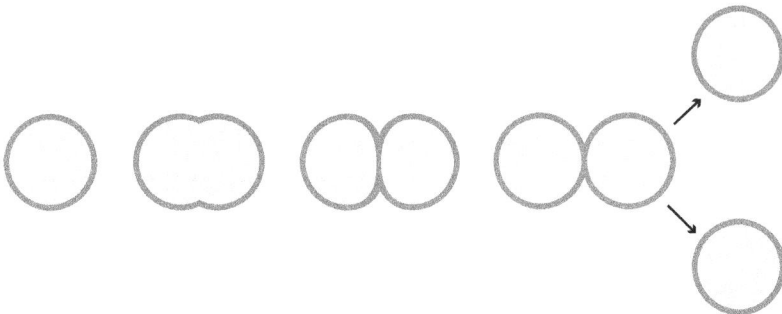

Fig. 4.1 Bacterial growth. A bacterial cell enlarges in size, then a wall separates the cell into two new cells exactly alike.

Bacteria Growth With Temperature

It can be seen on the graph that at 32° F (0° C) bacteria needs as much as 38 hours to divide in two. That also means that if our piece of meat had a certain amount of bacteria on its surface, after 38 hours of lying in a refrigerator the amount of bacteria will double. If we move this meat from the refrigerator to a room having a temperature of 80° F (26.5° C), the bacteria will divide every hour (12 times faster). At 90° F they will be dividing every 30 minutes.

Someone might say: why do I have to bother with all this bacteria stuff, I am going to kill them anyhow. Well, this is not entirely true, as *thermophillic* bacteria can survive high temperatures and might spoil the food if certain procedures are not observed. Secondly, if we let spoilage bacteria multiply, for example keeping meat at room temperature for too long, they will spoil the food. The food may not exhibit odor or slime yet, but its flavor is already affected. A thermal process will kill them but the fact remains they have already done some damage.

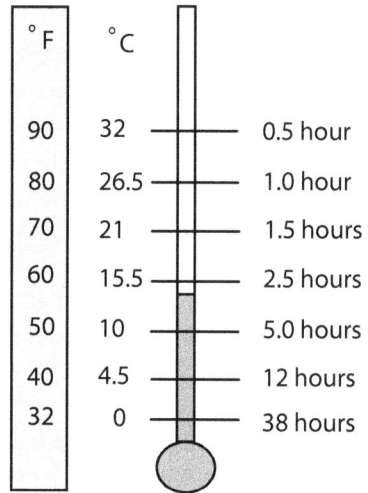

°F	°C	
90	32	0.5 hour
80	26.5	1.0 hour
70	21	1.5 hours
60	15.5	2.5 hours
50	10	5.0 hours
40	4.5	12 hours
32	0	38 hours

Fig. 4.2 Bacteria growth with temperature.

Food Spoilage Bacteria

How Do Bacteria Spoil Food? Microorganisms, like all living creatures must eat. Spoilage bacteria break down meat proteins and fats causing food to deteriorate and develop unpleasant odors, tastes, and textures. Fruits and vegetables get mushy or slimy and meat develops a bad odor. They don't use the toilet, but they excrete the waste which we end up eating. This is the unpleasant tasting "spoilage." Most people would not eat spoiled food. If they did, however, they probably would not get seriously sick. Bacteria such as *Pseudomonas spp.* or *Brochotrix thermosphacta* cause slime, discoloration and odors but do not produce toxins. There are different spoilage bacteria and each reproduces at specific temperatures. Some can grow at low temperatures in the refrigerator or freezer. Others grow well at room temperature and in the "Danger Zone" (40-140° F, 4-60° C). Most spoilage bacteria are easily killed by the temperature of boiling water, 212° F, 100° C.

There are, however, heat loving *thermophillic* bacteria which are so heat resistant that their spores can survive long exposure to the temperatures of 250° F, 121° C. They like to grow at 122-150° F, 50-66° C and this can create storage problems. If bacterial spores survive thermal treatment, they might find favorable conditions to grow upon slow cooling (122-150° F, 50-66° C). For that reason, commercial producers cool containers as fast as possible to about 95° F, 35° C. Very high storage temperatures, for example in tropical countries, will also create favorable conditions for their growth. To completely eliminate *thermophillic* bacteria the thermal process must be performed at temperatures even higher than 250° F, 121° C. A point must be made here that *thermophillic* bacteria may spoil the food, but *they do not produce toxins* and do not affect food safety. They are not classified as a pathogenic type.

Beneficial Bacteria

Without beneficial bacteria it would not be possible to make fermented sausages, sauerkraut, yogurt or cheeses. They are naturally occurring in the meat but in many cases they are added into the meat in the form of starter cultures. There are two classes of beneficial (friendly) bacteria:

- Lactic acid producing bacteria - *Lactobacillus*, *Pediococcus*.
- Color and flavor forming bacteria - *Staphylococcus*, *Kocuria* (previously known as *Micrococcus*).

They are easy to kill and most will die below the temperature of boiling water; they are of little concern when canning meats.

Pathogenic Bacteria

Pathogenic bacteria cause illness. They grow rapidly in the "Danger Zone" - the temperatures between 40 and 140° F - and *do not generally affect the taste, smell, or appearance of food.* Most pathogenic bacteria, including *Salmonella*, *E.coli 0157:H7*, *Listeria monocytogenes*, and *Campylobacter,* can be fairly easily destroyed using a mild cooking process.

Safety of Canned Products - It Is All About *Clostridium Botulinum*

Since the dawn of civilization, man has dealt with food poisoning. It has led to a number of deaths, but in most cases it was blamed on natural causes. There are very few historical sources and documents on the subject prior to the 19th century. In the 10th century Emperor Leo VI of Byzantium prohibited the manufacturing of blood sausage. At the end of the 18th century, there were documented outbreaks of "sausage poisoning" in Wurttemberg, Southern Germany. The biggest one occurred in 1793 in Wildbad where 13 people became ill (6 of whom died) after eating a locally produced blood sausage.

The above incident motivated the German poet and district medical officer Justinus Kerner (1786-1862) to investigate the problem. Although he did not succeed in discovering the bacteria that caused it, he was the first to publish detailed and complete descriptions of food poisoning between 1817-1822. He described 230 cases, most of which were linked to the consumption of sausages. He called it "sausage" or "fatty" poison. In time it became known as "botulism" after "botulus", the Latin word for sausage.

Eighty years after Kerner's work, in 1895, a botulism outbreak affected 34 people. After a funeral dinner in the small Belgian village of Ellezelles, a group of local musicians consumed smoked ham. That led to the discovery of the pathogen *Clostridium botulinum* by Emile Pierre van Ermengem, Professor of bacteriology at the University of Ghent who investigated the incident. Van Ermengen discovered that botulism was intoxication, not infection, and that the toxin was produced by a spore-forming obligate anaerobic bacterium, "*Clostridium botulinum.*" The term *Clostridium* means that the organism is able to grow in the absence of air and is a sporeformer. Home canned foods are the main source of botulinum food poisoning, however, two hundred years ago the canning process was still being developed.

Now we finally know that the biggest enemy of canned foods is *Clostridium botulinum*, a dangerous heat resistant microorganism which *does not need oxygen to grow*. If a canned product does not receive proper heat treatment, there is an increased risk that *Cl. botulinum* could survive and *produce toxin* within a container. The toxin attacks the nervous system and one millionth of a gram will kill a person. This means that 1 kg (2.2 lb) could kill 1 billion people on earth, clearly the strongest poison known to man.

Fortunately, we are seldom exposed to *Cl. botulinum* bacteria in their "vegetative" (growing) phase when they produce toxin. Yet, we find those bacteria in the form of bacterial spores in water and soil everywhere. Whenever a spore forming bacteria feels threatened, it will immediately build a protective wall around itself in the form of a cocoon. This shell is built within a few hours.

Fig. 4.3 Bacterial spore.

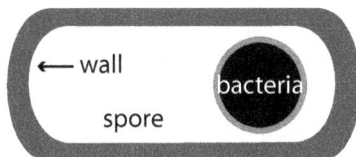

←— wall

bacteria

spore

Because *Cl.botulinum* hate oxygen, the air which is present in soil and water threatens them. *Cl.botulinum* bacteria immediately envelop themselves within a protective shell. They do not multiply, they just patiently remain inside waiting for more favorable conditions. Similarly to plant seeds, they can survive harmlessly in soil and water for many years. Then when the opportunity arises, they emerge from their shells and become vegetative bacteria (actively growing). During this growing stage they produce toxin. It is the toxin that kills, not the bacteria.

Where most bacteria can be killed at 160° F, 72° C, *Cl. botulinum* is protected inside of the spore and will survive the temperature of boiling water (212° F, 100° C) for 5 hours. Processing meat for so long will result in a poor texture, flavor and color. For this reason low acid foods must be processed in a pressure canner at 240° F, 116° C as this temperature will kill botulinum spores in about 2 minutes. If spores are not completely killed in canned foods, vegetative microorganisms will grow from the spores as soon as conditions are favorable again. Vegetative cells will multiply rapidly and may produce a deadly toxin within 3 to 4 days of growth in an environment consisting of:

- A moist, low-acid food.
- A temperature between 40° and 120° F (4-40° C).
- Less than 2% oxygen.

All of the above conditions are met in canned foods. Any surviving microorganisms can either spoil the canned food or produce toxins which cause food poisoning.

It is of absolute importance that food manufacturers implement policies that would prevent such an occurrence. And they do, however, home canners are not aware of the danger and often use procedures which are correct for making jams, but are completely wrong for processing meats or vegetables.

Botulinum spores are on most fresh food surfaces. Most bacteria, yeasts, and molds are difficult to remove from food surfaces. Washing fresh food reduces their numbers only slightly. Peeling root crops, underground stem crops, and tomatoes reduces their numbers greatly.

It would be ideal to apply very high temperatures which would eliminate all microorganisms, once and for all. Unfortunately, such a heat treatment will degrade the quality of some foods and will lower their nutritional value. A compromise is reached to keep the sterilization temperatures high enough to ensure safety of the products and low enough to produce quality products. The concept of heat treatment is introduced which is a combination of two components:

- Heating temperature.

- Heating time.

The same amount of heat treatment can be obtained when using lower temperature supported by a longer heating time *OR* higher temperature supported by a shorter heating time. It is easier to kill bacterial spores when the canning temperature is increased. The reference temperature of 250° F, 121° C and reference time of one minute is chosen. The amount of heat which is delivered at 250° F, 121° C, during one minute is defined as F-value-1. F-value for killing *Cl. botulinum* spores is 2.52, which means that the spores are deactivated when submitted to 2.52 minutes of heating time at 250° F, 121° C. This F-value of 2.5 for *C. botulinum* spores is known as *"botulinum cook."*

Time and Temperature Needed to Deactivate *Cl. botulinum* Spores		
214° F, 101° C	232° F, 111° C	250° F, 121° C
250 min	25 min	2.5 min

We ensure food safety with:

- Pasteurization, 149-203° F, 65-95° C. Pasteurization kills pathogenic bacteria and is adequate for foods that would be refrigerated. Typically, high acid foods or acidified foods (added vinegar, lemon juice) with a pH < 4.5 are pasteurized.

- Sterilization, 221-266° F, 105-130° C. Sterilization method makes indefinite product life at ambient temperatures. Typically, low acid foods with a pH > 4.5 are sterilized.

Blanching also helps, but the vital controls are the method of canning and making sure the recommended heating temperatures and times are used. The processing times in USDA guides ensure destruction of the largest expected number of heat-resistant microorganisms. Properly sterilized canned food will be free of spoilage if lids are sealed and jars are stored below 95° F (35° C). Storing jars at 50 - 70° F, 10 - 21° C enhances the retention of quality.

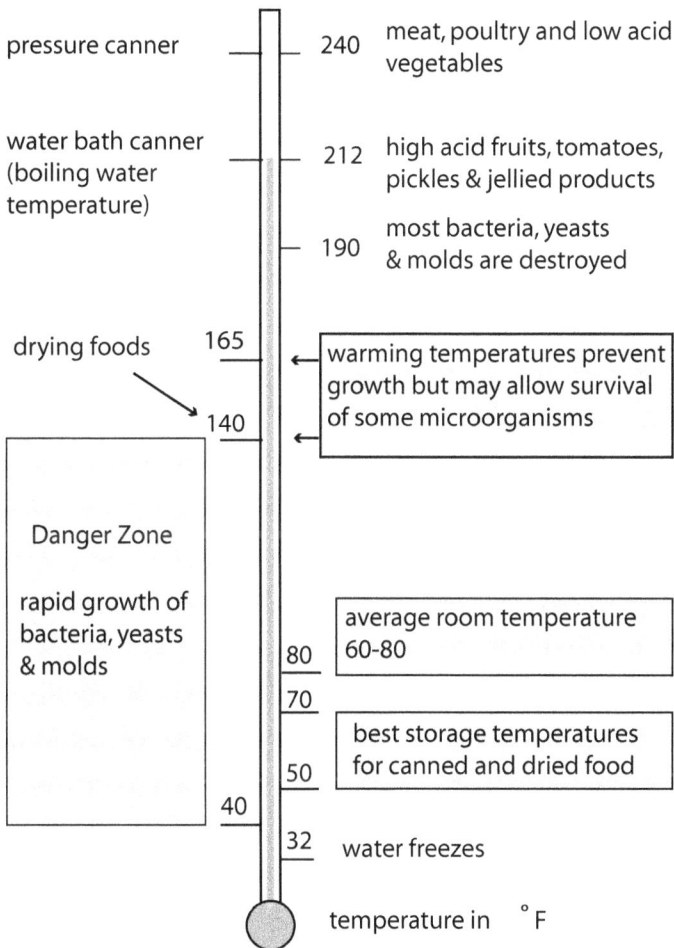

Fig. 4.4 Temperatures for food preservation.

Why Canned Foods Offer Favorable Conditions for the Growth of *Cl. botulinum*?

Cl.botulinum do not grow in the presence of air, so we are not at risk when cooking meats with usual cooking methods. Bacterial spores do not die, however, they remain in a dormant stage, enveloped in their protective shells until conditions become favorable for their growth. This happens when there is no more oxygen (air) and this is exactly what happens during canning when most of the available air is exhausted from the container. In order to form a strong vacuum inside of the container, as much air as possible is removed. This creates a stronger seal and leaves room for the expansion of gasses. Without a vacuum, the cans will buckle and the contents of a jar may boil over through the seal. In a hermetically sealed container bacterial spores find the right temperature, plenty of moisture and nutrients, and the absence of air. They can leave the spore and germinate. They start growing and produce toxin.

Control of *Cl.botulinum*

There are two ways of controlling *Cl.botulinum*:

- Killing spores.
- Preventing spores from germinating and growing.

Spores of *Cl.botulinum* are present in both acidic and low-acid foods. In low acid-foods such as meats and vegetables, *Cl.botulinum* spores can only be killed at 240° F, 116° C or higher. The high acidity (pH < 4.6), however, prevents botulinum bacteria from leaving the spores. This prevents them from germinating and growing, and of course no toxin is produced. For this reason *high-acid foods* such as fruits or juices can be processed at 212° F, 100° C as this temperature will kill all bacteria in vegetative form and bacterial spores are prevented from germinating by high acidity.

The growth of *Cl.botulinum* is inhibited at 10% salt concentration which is equivalent to a water activity of around 0.93. Obviously, such high salt percentages will not be tolerated by a consumer.

Inactivating Toxin

The toxin is not heat resistant; it can be inactivated by boiling in water (212° F, 100° C) for 10 minutes. Old canning manuals often asked for boiling home canned meats and vegetables for 10-15 minutes in an *open* vessel. This procedure was meant to destroy any bacteria or toxins that might have survived the incorrect canning process.

Food Acidity and Processing Methods

The main objective of thermal processing is control of *Cl.botulinum* bacteria. Whether food should be processed in a pressure canner or boiling-water canner depends on the acidity of the food. The term "pH" is a measure of acidity; the lower its value, the more acidic the food. Bacteria hate acidity, this fact works to our advantage. Acidity may be natural, as in most fruits, or added, as in pickled food. The acidity level in foods can be increased by adding lemon juice, citric acid, or vinegar. All bacteria have their own preferred acidity level for growth, generally around neutral pH (7.0). Bacteria will not grow when the pH is below the minimum or above the maximum limit for a particular bacteria strain. As the pH of foods can be adjusted, this procedure becomes a potent weapon for the control of *Cl.botulinum*.

The thermal resistance of microorganisms decreases as the pH of their medium is lowered. As explained earlier, most bacteria, particularly *Cl. botulinum*, will not grow below pH 4.6. Therefore acidic foods having pH below 4.6 do not require as severe heat treatment as those with pH above 4.6 (low acid) to achieve microbiological safety.

The pH value of 4.6 is the division between high acid foods and low acid foods. Low-acid foods have pH values higher than 4.6. They include red meats, seafood, poultry, milk, and all fresh vegetables except for most tomatoes.

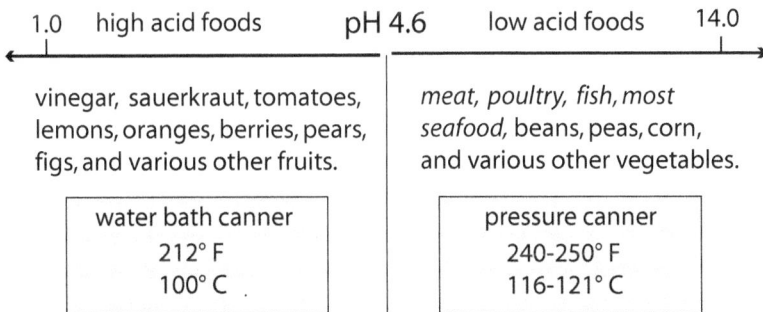

1.0 high acid foods	pH 4.6	low acid foods 14.0
vinegar, sauerkraut, tomatoes, lemons, oranges, berries, pears, figs, and various other fruits.		*meat, poultry, fish,* most *seafood,* beans, peas, corn, and various other vegetables.
water bath canner 212° F 100° C		pressure canner 240-250° F 116-121° C

Fig 4.5 Processing methods for low and high acid foods.

Most mixtures of low-acid and acidic foods also have pH values above 4.6 unless their recipes include enough lemon juice, citric acid, or vinegar to make them acidic foods. Acidic foods have a pH of 4.6 or lower. They include fruits, pickles, sauerkraut, jams, jellies, marmalades, and fruit butters.

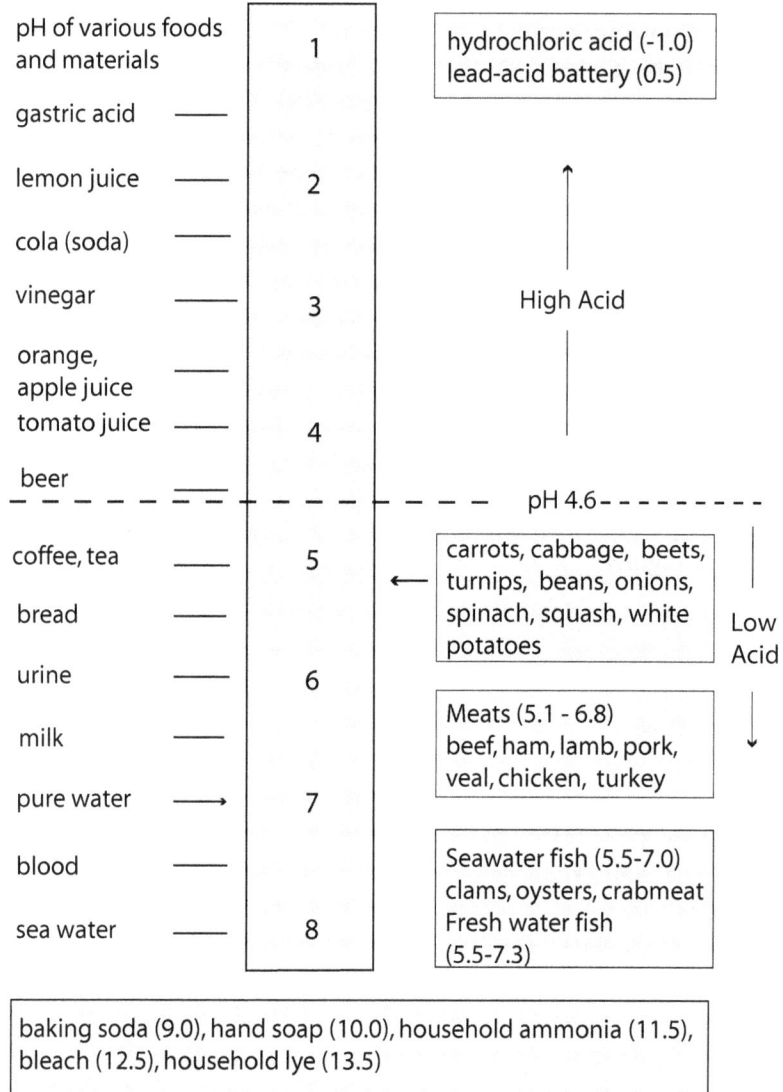

pH of various foods and materials		hydrochloric acid (-1.0) lead-acid battery (0.5)
	1	
gastric acid ——		
lemon juice ——	2	
cola (soda) ——		↑
vinegar ——	3	High Acid
orange, apple juice ——		
tomato juice ——	4	
beer ——		

– – – – – – – – – – – – – – – – pH 4.6 – – – – – – – – –

coffee, tea ——	5	← carrots, cabbage, beets, turnips, beans, onions, spinach, squash, white potatoes	Low Acid
bread ——			
urine ——	6	Meats (5.1 - 6.8) beef, ham, lamb, pork, veal, chicken, turkey	↓
milk ——			
pure water ——→	7		
blood ——		Seawater fish (5.5-7.0) clams, oysters, crabmeat Fresh water fish (5.5-7.3)	
sea water ——	8		

baking soda (9.0), hand soap (10.0), household ammonia (11.5), bleach (12.5), household lye (13.5)

Fig. 4.6 pH of various foods and materials. A lower pH value indicates higher acidity.

Although tomatoes generally fall on the pH dividing line at 4.6, there are varieties that have a lower pH level and there are varieties which also have a higher pH level. Tomatoes are usually considered an acidic food. Figs also have pH values slightly above 4.6. *Therefore, if they are to be canned as acidic foods, these products must be acidified to a pH of 4.6 or lower with lemon juice, citric acid or vinegar.* Properly acidified tomatoes and figs are acidic foods and can be safely processed in a boiling-water canner. Changing acidity levels, however, requires a good understanding of food technology and should be left to professionals. In addition, not all foods will appeal to a consumer when their flavor is overly acidic.

pH Meters

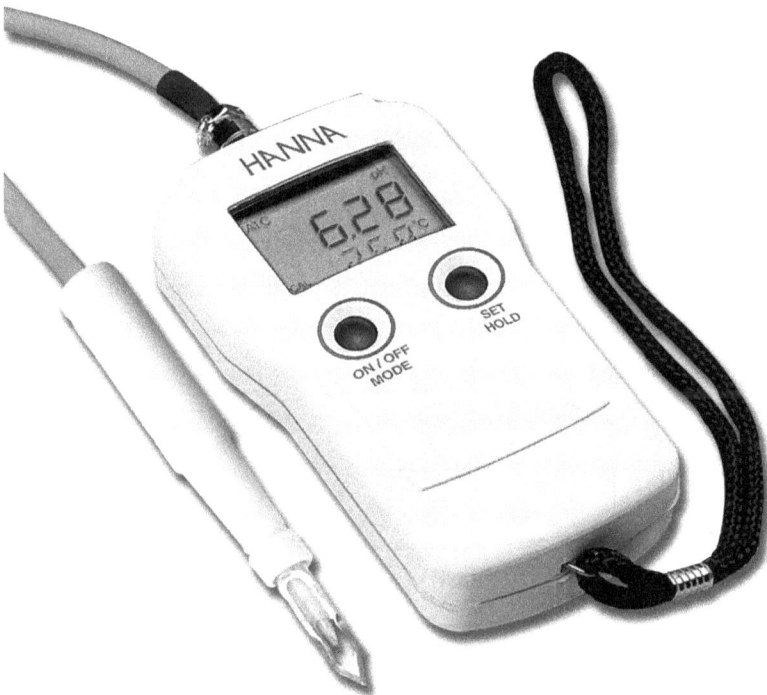

Photo 4.1 HI 99163 Portable meat pH meter.

Photo courtesy Hanna Instruments http://www.hannainst.com

There are many companies making pH measuring equipment. Large bench models are expensive and will be used by the professionals who establish thermal processes for canned foods. We have been using the Hanna Instruments HI 99163 Meat pH Meter with much satisfaction for making fermented sausages. The instrument is of great value for the pH analysis of meat products. This pH meter is simple to use with only two buttons. The replaceable penetration blade allows the user to measure not only the surface, but also the internal pH of the meat. The unit is accurate to within 0.02 pH and the reading is obtained within seconds.

Control of Bacteria by Water Activity

In order to remain alive the microorganisms need nutrients and moisture. *How do bacteria eat?* They do not have mouths so they have to absorb food differently. They have to dissolve food in water first and then the food can be absorbed. Bacteria are like a sponge, they absorb food through the wall, but the food must be in a form of a solution.

Imagine some sugar, flour or bread crumbs spilled on the table. Place a dry sponge on top of the sugar and you will see that the sponge will not pick up any of the ingredients. Pour some water over the ingredients and repeat the sponge procedure. The sponge will absorb the solution without any difficulty.

To summarize: when water is eliminated, bacteria cannot eat.

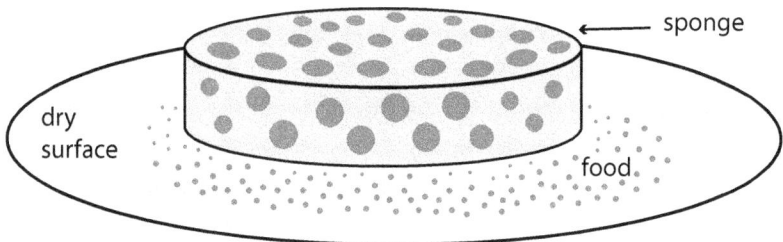

Fig. 4.7 Sponge will not pick up food particles from a dry surface.

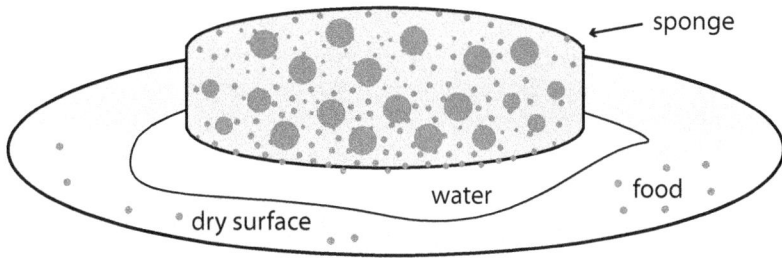

Fig. 4.8 Sponge easily picks up food particles that dissolved in water.

All microorganisms need water and the amount of water available to them is defined as water activity. Water activity (Aw) is an indication of how tightly water is "bound" inside of a product. It does not say how much water is there, but how much water is *available* to support the growth of bacteria, yeasts or molds. If you squeeze steak, the juice will be released. This juice is free water available to bacteria. Bound water remains inside although some of it will be removed during cooking. Adding salt or sugar "binds" some of this free water inside of the product and lowers the amount of available water to bacteria which inhibits their growth. The most practical approach for lowering water activity is drying, although it is a slow process which must be carefully monitored, otherwise it may backfire and ruin the product. A simple scale is used to classify foods by their water activity and it starts at 0 (bone dry) and ends on 1 (pure water).

Below certain Aw levels, microbes can not grow. USDA guidelines state:

"A potentially hazardous food does not include . . . a food with a water activity value of 0.85 or less."

Meats, dried fruits and vegetables were preserved throughout our history and the technology was based on simple techniques of salting and drying.

Water activity (Aw) of some foods	
Pure water	1.00
Fresh meat & fish	0.99
Bread	0.99
Salami	0.87
Aged cheese	0.85
Jams & jellies	0.80
Plum pudding	0.80
Dried fruits	0.60
Biscuits	0.30
Milk powder	0.20
Instant coffee	0.20
Bone dry	0.00

It was also discovered that the addition of sugar would preserve foods such as candies and jellies. Both factors contribute to lowering the water activity of the meat. Freshly minced meat possesses a very high water activity level around 0.99, which is a breeding ground for bacteria. Adding salt to meat drops this value immediately to 0.96-0.98 (depending on the amount of salt), and this already creates a hurdle against the growth of bacteria. This may be hard to comprehend as we know that water does not suddenly evaporate when salt is added to meat. Well, this is where the concept of water activity becomes useful.

Although the addition of salt to meat does not force water to evaporate, it does something similar: it immobilizes free water and prevents it from reacting with anything else, including bacteria. It is like stealing food from the bacteria, the salt locks up the water creating less favorable conditions for bacteria to grow and prosper. As we add more salt, more free water is immobilized but a compromise must be reached, as adding too much salt will make the product unpalatable. It may also impede the growth of friendly bacteria, the ones which work with us to ferment the sausage. The same happens when we freeze meat though we never think of it. Frozen water takes the shape of solid ice crystals and is not free anymore. Water exists in meat as:

- Bound (restricted or immobilized water) - structurally associated with meat proteins, membranes and connective tissues. This water (3-5% of total water) can only be removed by high heat and is not available for microbial activities.

- Free or bulk water - held only by weak forces such as capillary action. This free water is available for microorganisms for growth.

Removing water content by drying food has been practiced for centuries. As the process proceeds, water starts to evaporate (water activity decreases), making meat stronger against spoilage and pathogenic bacteria. There eventually comes a point where there are no bacteria present and the meat is microbiologically stable. It will not spoil, as long as it is kept at low temperatures and at low humidity levels. If the temperature and humidity go up, new bacteria will establish a colony on the surface and will start moving towards the inside of the sausage. The mold will then immediately appear on the surface. Microbial spoilage is a frequent cause of spoiled foods or defective containers.

Microbial spoilage may result from:

- Holding containers too long before processing.
- Contamination after processing.
- Inadequate thermal process.
- Inadequate cooling of the containers or storage at high temperatures.
- Spoilage due to survival of acid-tolerant spores (*flat sour*).

Minimum Aw requirements for microorganism growth	
Molds	0.75
Staphylococcus aureus	0.85
Yeasts	0.88
Listeria	0.92
Salmonella	0.93
Cl.botulinum	0.93
E.coli	0.95
Campylobacter	0.98

Water Activity Meters

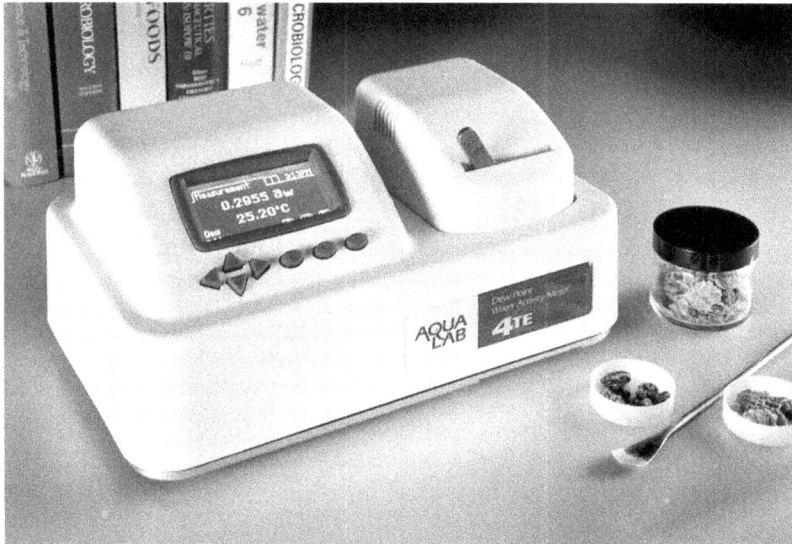

Photo 4.2 AquaLab 4TE benchtop water activity meter.
Photo courtesy Decagon Devices Inc., www.decagon.com

Decagon Devices Inc. makes top of the line water activity meters, like the benchtop model depicted above, which is accurate to ± 0.003 Aw. They also make a highly popular portable model that is a favorite with meat inspectors as it fits into the pocket.

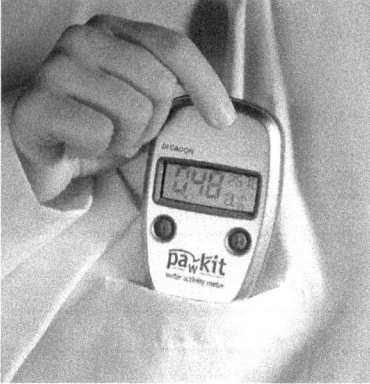

Photo 4.3 Pawkit water activity meter.

Resolution: plus, minus 0.01 Aw.
Accuracy: plus, minus 0.02 Aw.
Case dimensions: 3.5" x 4."
Weight: 115 g (4 oz).

Photo courtesy Decagon Devices Inc., www.decagon.com

Summary

Botulinum spores are very hard to destroy at boiling-water temperatures; the higher the canner temperature, the more easily they are destroyed. Therefore, *all low-acid foods should be sterilized at temperatures of 240° to 250°F, 116 - 121° C,* attainable with pressure canners operated at 10 to 15 PSIG (PSI). PSIG means pounds per square inch of pressure as measured by gauge. *At temperatures of 240° to 250°F, 116 - 121° C, the time needed to destroy bacteria in low-acid canned food ranges from 20 to 100 minutes.* The exact time depends on the kind of food being canned, the way it is packed into jars, and the size of the jars. The time needed to safely process *low-acid* foods in a *boiling-water canner* ranges from 7 to 11 hours; the time needed to process *acid* foods in *boiling water* varies from 5 to 85 minutes.

The USDA recommends that most meats are cooked to 160° F, 72° C, internal temperature so a question may be given why canned meats must be processed to 240-250° F, 116-121° C. Well, cooking meats to 160° F, 72° C internal meat temperature kills ordinary bacteria so the meat is considered safe to eat, but be aware that bacterial spores will survive. Because they are in continuous contact with air, they will not germinate and will not produce toxin. You might eat them, but that will not create danger, after all we eat bacteria in sauerkraut, salami, cheese and yogurt all the time.

Chapter 5

Principles of Canning Low-Acid Foods

Meats, poultry, fish and vegetables are classified as low-acid foods (pH > 4.6) and must be processed until a condition of *"commercial sterility"* is achieved. Commercial sterility is defined as the condition obtained by the application of heat alone or in combination with other treatments to render the product free of microorganisms capable of growing in the product at normal *non-refrigerated* conditions. The product is safe to eat because the pathogenic microorganisms are either destroyed or inactivated to the extent that they pose no health risk. The product will remain shelf stable as long as the container is perfectly sealed, which is what prevents the entry of any microorganisms from the outside. Since spoilage bacteria are easily killed during heat treatment, the product has an almost unlimited shelf life as long as it is stored at proper conditions. After a year its quality, however, will start to deteriorate but the product will remain safe to eat. There are two methods of canning employed by the industry:

- Conventional canning. Home made canning fits into this description.

- Asceptic canning. Foods and equipment are both prepared separately in sterile conditions, then combined together in a sterile room. This method cannot be employed at home and will not be described in this book.

Almost all of home canning is done in glass jars, but *this does not mean that jars are superior to cans*. If they were, commercial manufacturers will use them instead of metal cans, won't they? The cans are used everywhere: the Armed Forces, Veterans Administration, school sunches, seedy samilies programs, hotels, restaurants, colleges, penal institutions, catering services, the list is endless.

The main reason why metal cans are less popular among home canners is the total lack of information on the subject. All books, manuals and guides that were written after 1950 do not mention canning in metal containers as a viable option for a hobbyist at all. The original work that was done by the USDA and other agencies culminated in 1946 with manuals on the subject of meat canning. Both methods, *glass jars and tin cans* were covered and *processing times were given for both.* Unfortunately, starting with the 1988 revision, the information on canning in tin cans was omitted from all USDA editions of the 539 Bulletin, which is largely a reprint of the AWI-110 original work from 1946.

Canning in glass jars is simpler and there is less room for error, so the USDA chose the easiest path for solving safety problems: *if we don't offer information, people will not be using metal cans.* It seems that the USDA has taken for granted that people are not capable of comprehending a few basic procedures. Well, many of us are willing to study the subject of canning in more detail in order to make it safe, even at home. Our hat to the University of Alaska in Fairbanks; the first institution on the Internet which decided that canning in metal cans is a suitable method for a home canner. Many home canners in British Columbia and in Alaska are canning game meat and fish in tin cans today.

Today, we can find any kind of food packed in metal cans; soups, stews, beans, chicken, noodles, fish, meat, fat, oysters, clams, the list is endless. Almost all fish is canned in metal containers. It is a sad fact that canning meat products in metal containers does not get any recognition from the USDA, at least for canning products at home, and hopefully this book will help people to understand that canning in metal containers is not rocket science.

It is confusing for a newcomer to select the right can. The next problem is to determine what type of sealer to buy that will seal a particular can. It is easy to end up with an expensive sealer which might be a good choice for sealing large cans with fruit, but a poor choice for canning meats. However, there are great little sealers that have been performing wonderfully for decades and they are a perfect choice for a home canner. The best example is the Ives-Way can sealer which has been made in the US for 60 years. It costs as little as any kitchen appliance, and it can seal a great variety of cans by switching inexpensive chuck adapters and spacers. The set up and adjustments are very easy and the telephone support is excellent.

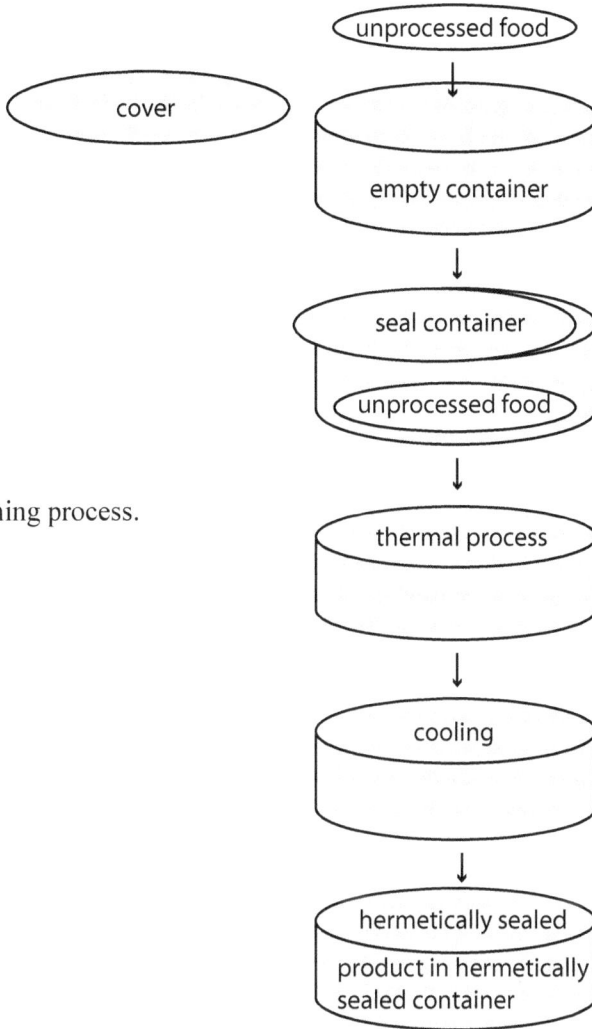

Fig. 5.1 Canning process.

The main steps in canning are:

1. Filling the product into the container.
2. Hermetically sealing the container.
3. Thermally processing the product and the container together.
4. Cooling.

Hermetically sealed container means a container that is designed and intended to be secure against the entry of microorganisms and therefore maintains the commercial sterility of its contents after processing.

1. Filling Containers

Commercial packers use containers of all shapes, sizes and materials. Glass, steel, aluminum, plastic-cardboard-aluminum combinations, plastic containers and all types of closures. Such containers are processed by specialized equipment that is not available to a hobbyist. A home canner will can his products in glass jars or in metal cans, both types of containers are described in detail in the Equipment section of this book. How the product is prepared and packed will influence the heat penetration of the container and will be taken under consideration by processing authority when evaluating the process schedule of a new recipe.

Meat Preparation

Meat intended for canning should not be permitted to freeze. If it does freeze, keep it frozen until canning time because thawed meat spoils very quickly. Wash meat, if necessary, but do not soak in water. To do so dissolves meat juices and renders meat stringy. Wiping off with a damp cloth is usually all that is needed.

Cut meat into convenient pieces for packing in jars. Small bones may be left on. They seem to improve the flavor and aid in heat penetration.

Precooking Meat

Meat precooked in water is more like boiled meat in texture and flavor while meat precooked in the oven resembles roasted meat. Precooking in water may be referred to as *parboiling*, as the fastest way to precook a large quantity of meat and is often used with chicken. Frying makes the meat harder and gives it a less desirable flavor. Pan broiling gives good results.

Glass jars - meats should be precooked in water or in the oven before being packed.

Metal cans - meats may be:

- precooked in water or in the oven and packed hot *OR*
- packed raw and precooked in the cans while they are being exhausted before being sealed. This method gives a better flavored product as it saves all meat juice in the can, but it requires more time and space.

Methods of Packing

Raw-Packing - is the practice of filling jars tightly with freshly prepared, but *unheated* food. Such foods, especially fruit, will float in the jars. The entrapped air in and around the food may cause discoloration within 2 to 3 months of storage. Raw-packing is more suitable for vegetables processed in a pressure canner.

Hot-Packing - is the practice of heating freshly prepared food to boiling, simmering from 2 to 5 minutes, and promptly filling jars loosely with the boiled food. Boiling *hot liquid* is added to the jars.

Photo. 5.1 Raw pack - *add very hot canning liquid or water to cover raw food*, but leave *headspace*. Fruits and vegetables packed raw should be packed *tightly* because they will shrink during processing. The exceptions are corn, lima beans, potatoes and peas that should be packed loosely because they expand during canning.

Photo. 5.2 Hot pack - raw foods are boiled 3 - 5 minutes in a saucepan, then poured into jars. This helps to remove air from food tissues, shrinks food, helps keep the food from floating in the jars, increases the vacuum in sealed jars, and improves shelf life. Foods packed hot should be packed *loosely*, as shrinkage has already taken place.

Canned meat retains flavor and color better during storage when the meat in each can is entirely covered with liquid. Water or drippings from the pan in which the meat was precooked should be added to the meat after it is packed into the jar or the can. Some juice will be released from meat precooked in cans during exhausting, if that is not enough to cover the meat, hot water should be added before sealing. Whether food has been hot-packed or raw-packed, the juice, syrup, or water to be added to the foods should also be *heated to boiling* before adding it to the jars. At first the color of hot-packed foods may appear no better than that of raw-packed foods, but within a short storage period, both color and flavor of hot-packed foods will be superior. Pre-shrinking food permits filling more food into each jar.

Photo 5.3 Removing air. Air bubbles are removed by running a plastic knife or spatula around the inside of the jar.

There should be enough liquid to fill in around the solid food in the jar and to cover the food. The food which is not covered by liquid tends to darken and develop off-flavors. A common example are peeled raw potatoes which will darken unless covered with water.

Glass Jars

Hot-pack. The meats are heated (precooked) before packing into the jars. Pack precooked meats loosely, fill with hot broth or hot water leaving 1 inch headspace.

Raw-pack. The raw meats are packed tightly into the jars all the way to the top. When processed in the canner, the meats will shrink and release juice.

Metal Cans

Hot-pack. The meats are precooked and packed into cans hot. Then they are filled with hot liquid (pan drippings, meat broth or hot water) leaving ¼ inch headspace. Seal and process at once.

Raw-pack. The meats are packed into the cans raw to within ½ inch of the top. Then they are heated (exhausted) to the temperature at the center of 170° F, 77° C. If needed, add more hot liquid to within ¼ inch of the top. Seal and process at once.

Notes:

Filled containers should be sealed as soon as possible when still hot and placed in hot water in the canner. Then they should be immediately processed.

The higher the filling temperature, the less pressure will be generated in the container by heating the contents. As a result a stronger vacuum forms in the container after thermal processing and cooling.

Salt does not preserve meat in canning and is added for flavoring. It can even be left out altogether. If used, place salt in the container before putting in the meat.

Headspace

The unfilled space above the food in a sealed container and below its lid is termed headspace. The amount of headspace required depends on the type of food being canned. For example, starchy foods tend to expand when heated and therefore require more headspace.

Headspace is of lesser importance in metal cans because the cans withstand the inside pressure quite well and allow the food to expand without spreading the seams. It is, however, accepted trade practice to reserve about 6% of the volume of the can for headspace. The meats packed in cans in home production are usually filled with hot broth or hot water leaving about ¼ inch headspace. Heating air is a slow process so any additional volume of air will adversely affect the heat transfer.

Leaving the specified amount of headspace in a jar is important to assure a vacuum seal. If too little headspace is present the food may expand and bubble out when air is being forced out from under the lid during processing. The bubbling food, especially fat, may leave a deposit on the rim of the jar or the seal of the lid and prevent the jar from sealing properly. If too much headspace is present, the food at the top is likely to discolor. Also, the jar may not seal properly because there will not be enough processing time to drive all the air out of the jar. And more air means more oxygen available to discolor the food and promote rancidity in fats.

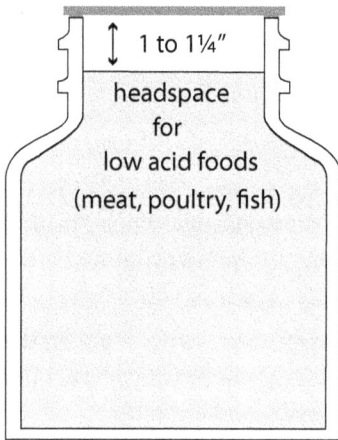

Fig. 5.2 Headspace.

Headspace is needed for the expansion of food as jars are processed and for forming vacuum upon cooling. The extent of the expansion is determined by the air content in the food and by the processing temperature. Air expands greatly when heated to high temperatures; the higher the temperature the greater the expansion. Foods expand less than air when heated.

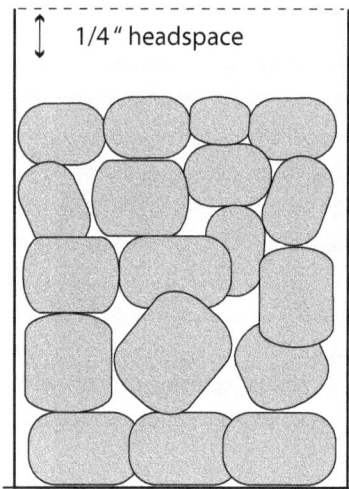

The headspace for most products processed in cans at 240-250° F, 116-121° C, should be no less than 6% and no more than 10%. The proper amount of headspace contributes to the formation of a vacuum inside a can and is needed to accommodate the expanding food and gasses.

Fig. 5.3 Headspace in a metal can. The maximum headspace for a 90% fill of No. 2 can (307 x 409) is ½".

Headspace in Glass Jars	
Meats, poultry and fish	1 - 1¼ inch
Headspace in Metal Cans	
Meats, poultry and fish	No. 2 can (307 x 409), ½ inch
	No. 3 can (404 x 414), ¾ inch
Meat products containing cereal should be packed with more headspace to allow for greater expansion.	

Vacuum

Vacuum is a measure of the extent to which air has been eliminated from the container. The amount of air that is left in the container after filling and the amount of vacuum are closely related.

Fig. 5.4 The container is filled with *hot food* and sealed. A little air remains inside after sealing which is an indication of a strong vacuum. The container is ready for thermal treatment.

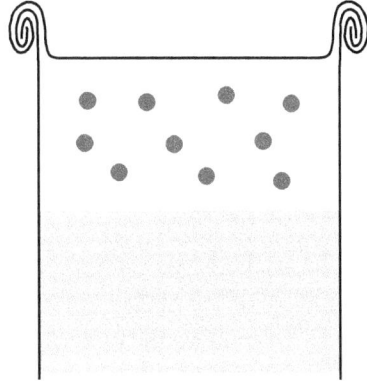

Fig. 5.5 Heat has been applied and the amount of air molecules remains the same. However, the molecules start to move faster and collide with the sides, the lid and each other. They start to exert pressure on the body of the container. As more heat is applied, the air molecules move even faster causing the pressure and temperature to increase.

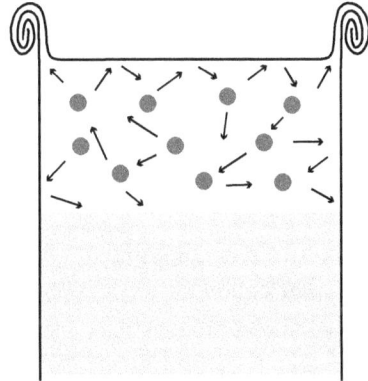

Fig. 5.6 The same container filled with *cold food* and sealed. As a result *more air* remains inside and the vacuum is weak. Heat is applied and the air molecules *start moving around building up the pressure*. The can contains plenty of molecules of air which have no place to go. They start to exert a lot of pressure on the lid and the seams of the can. This may weaken the seam and even deform the can.

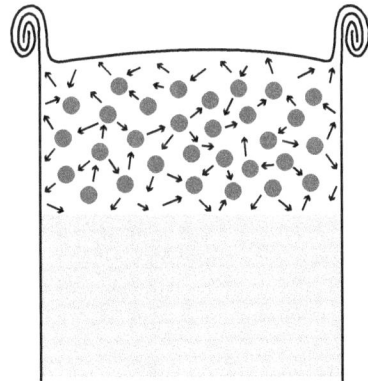

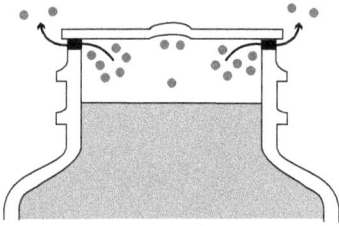

Fig. 5.7 Glass jars do not face seaming or cover deforming problems as the accumulating pressure can escape through the still soft sealing compound.

A strong vacuum provides the following benefits:

- It reduces stress to the can and its seams during thermal processing.
- It maintains the can ends or jar lids in a concave position giving a visual indication to the conditions of the container.
- It reduces the quantity of oxygen in the container. Fats are not going rancid, the food maintains its quality longer.

In food containers a vacuum is produced by the following methods:

- Thermal exhaust
- Steam displacement
- Mechanical action

Exhausting

Exhausting is allowing air or similar gases to escape from the food. In a sealed container oxygen is undesirable, whether it is released from food cells or is present in the form of entrapped air. Oxygen may react with the food and the interior of the can and affect the quality and nutritive value of the canned food. Other gases, for example, carbon dioxide, should also be exhausted as much as possible. The gases may place undue strain on the container during the heat process since gases expand. This will be more of a concern in metal cans, where the gases will be hermetically trapped and have no means to escape.

Thermal Exhaust. This is a typical home production method.

Cans: contents of the container are heated to 170° F, 77° C, prior to sealing the container. This is the temperature needed to exhaust air properly so that a strong vacuum forms inside the can upon cooling.

Jars: The same effect is produced by filling jars with hot food, and adding boiling water, broth, syrup or brine to the container.

Air bubbles may be trapped inside the jar and will raise to the top during processing, increasing headspace. This may adversely affect the closure of the jar. Run a plastic utensil (knife, spatula) around the jar, moving it up and down, so that any trapped air is released.

In commercial applications exhausting is accomplished by:

Steam Displacement. Steam is introduced into the headspace where it forces air out. When the container cools down, the steam condenses and a vacuum is produced. Filled with food, open containers are passed through an "exhaust box" in which steam is used to expand the food by heat and expel air and other gases.

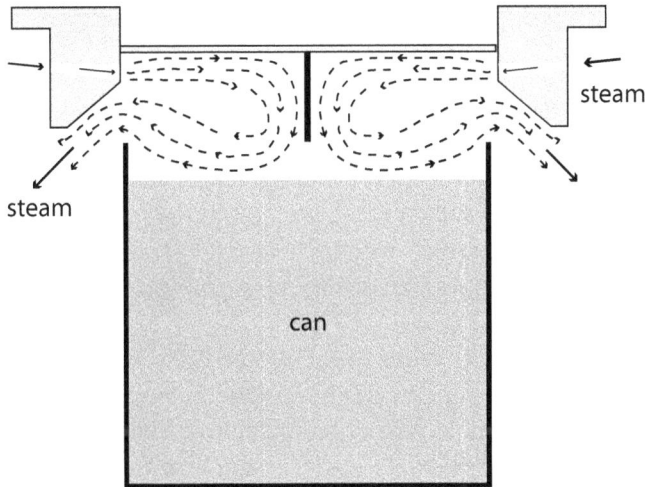

Fig. 5.8 Obtaining a vacuum by injecting steam into headspace. The steam pushes the air out, then the can is immediately sealed. When the steam condenses, a vacuum is formed.

Mechanical. A commercial method. A portion of the air in the container headspace is removed with a pump. Regardless of the exhausting method used, the container must be immediately sealed while it is still hot.

Note: After exhausting the metal cans should be processed at once, while still hot. Cans are never sealed cold.

2. Sealing the Container

The sealing operation is one of the most important in the canning process. The heat destroys spoiling, pathogenic bacteria and bacterial spores which may be present in the food; the seal on the container prevents re-entry of such organisms from the outside. The seal is accomplished by screwing the lid on glass jars and bending the curl of the lid with the flange of the can body together into what is known as a "double seam." However, in both cases a compound sealant, which is attached to the lid, will complete the seal.

Sealing glass jars is a simple process and the closure is easily checked for any possible defects. Sealing cans is a more difficult process that requires a sealing machine and more knowledge on the part of the operator. A finish seam may look fine, yet it may be faulty inside which can be verified only by tearing the can apart.

3. Thermal Process

Low-acid foods such as meat, poultry, fish and vegetables *must* be processed at 240-250° F, 116-121° C temperatures to kill *Clostridium botulinum* spores. Low-acid foods means any foods, other than alcoholic beverages, with a finished equilibrium pH greater than 4.6 and a water activity (Aw) greater than 0.85. Tomatoes and tomato products having a finished equilibrium pH less than 4.7 are not classified as low-acid foods.

In an open kettle at sea level, water boils at 212° F, 100° C which is insufficient for destroying *Cl.botulinum* spores. At increased pressures water boils at higher temperatures, hence the need for a pressure canner. Developing the processing time for a food product to be canned is a complex and expensive process. It depends on:

- The number and the type of bacteria present in the food.
- The rate of heat transfer through a food in a given container.
- The temperature and the time of the heating.
- The pH of the food.

In the past, the USDA working with the National Canners Association conducted thermal processing studies for different foods. To determine a processing time for each food, bacteria which were more heat resistant than *Clostridium botulinum* were introduced into sample jars.

The test organism was putrefactive anaerobe No. 3679, isolated by Cameron in 1927 in the laboratories of the National Canners Association. The spores of the organism exhibit a resistance to heat almost twice the maximum resistance reported for *Cl. botulinum* under the same conditions.

The jars were heated for different times, then they were held and tested for spoilage. At each temperature and processing time the reading was taken when bacteria started to die. The results of the tests were plotted on a graph where a single curve called Thermal Death Time, displays the findings. The calculated times were verified by actual tests with jars inoculated with known quantities of bacteria. A margin of safety was introduced and the processing time and temperature for a particular food in a given jar was established. The United States Department of Agriculture Information *Bulletin No. 539* lists detailed processing times for a great variety of foods.

Processing Authority

Processing times that commercial producers use are much shorter in order not to overcook the food and preserve the best texture and color. Many factors are involved in designing a new recipe such as the properties of raw material, the amount of sugar, the percentage of fat, thickening materials (starch, flour), acidifying agents (lactic acid, citric acid, lemon juice) and more. Thermal processes for canned foods are established by persons who have an expert knowledge in this field. In addition they must possess advanced testing equipment which will allow them to conduct heat penetration tests and /or inoculated pack studies. For example, fish can be packed in brine, sauce or oil and each method will affect the heating characteristics of the product. A product packed in oil needs longer processing times as bacterial spores receive extra protection from a thin film of oil that encapsulates them.

A process is established for a particular food, formulation, thermal processing details, container type and size. Any changes to the formulation, process or container size must be evaluated and approved by a processing authority and not by the packer.

The table below demonstrates the importance of time-temperature combinations for killing bacteria. The information comes from Putra University, Malaysia: From thermal death curves, the following time/ temperature treatments yield the same microbe killing effect:

° F	212	219	230	240	244	250	255	260
° C	100	104	110	116	118	121	124	127
min.	330	150	36	10	5.27	2.78	1.45	0.78

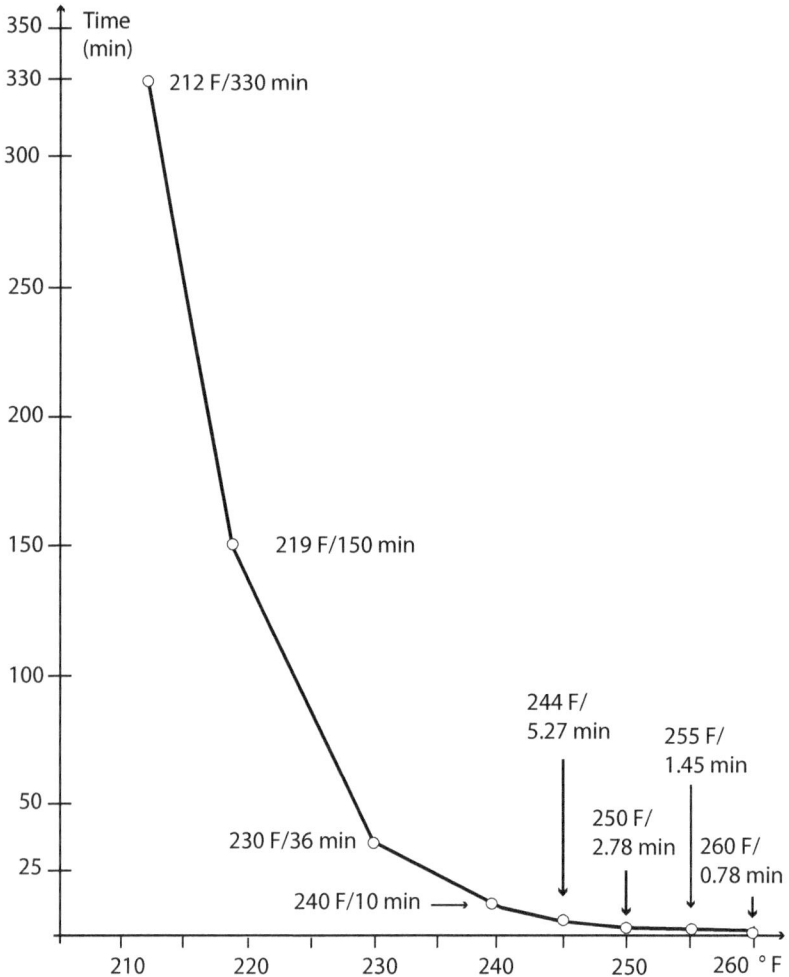

Fig. 5.9 Time-temperature relationship for killing bacterial spores.

Process Adjustments at High Altitudes

Using the process time for canning food at sea level may result in spoilage if you live at altitudes of 1,000 feet or more.

Water boils at lower temperatures as altitude increases. Increasing the process time or canner pressure compensates for lower boiling temperatures. Therefore, when you use the guides, select the proper time and pressure for the altitude where you are canning.

feet	temp.
10,000	194 F
8,000	197 F
6,000	201 F
4,000	204 F
2,000	208 F
sea level	212 F

Fig. 5.10 Water boiling temperature at different altitudes.

The following table can be used for pressure and temperature conversion for canners that use the metric system.

\multicolumn{5}{Boiling water temperature and associated pressure}				
° F	° C	Pounds per square inch (psi)	Atmosphere (atm)	Bar
212	100	14.69	1.00	1.01
221	105	17.57	1.19	1.21
229	108	19.45	1.32	1.34
228	109	20.11	1.36	1.38
230	110	20.79	1.41	1.43
239	115	24.48	1.66	1.68
240	115.5	24.48	1.69	1.71
248	120	28.71	1.95	1.97
250	121.1	29.62	2.01	2.04
260	127	35.62	2.42	2.45

A weighted gauge comes in three pressure settings: 5, 10 and 15 lbs. After the canner is being vented, the pressure inside is 14.69 psi and the temperature is 212° F, 100° C.

Adding a 5 lb weighted gauge increases pressure to 14.69 + 5 = 19.69 lb of pressure which corresponds to about 227° F, 108.5° C. Adding a 10 lb weighted gauge increases pressure to 14.69 + 10 = 24.69 lb of pressure which corresponds to about 240° F/116° C. Adding a 15 lb weighted gauge increases pressure to: 14.69 + 15 = 29.69 psi. This corresponds to 250° F, 121° C.

Factors Influencing Heat Transfer

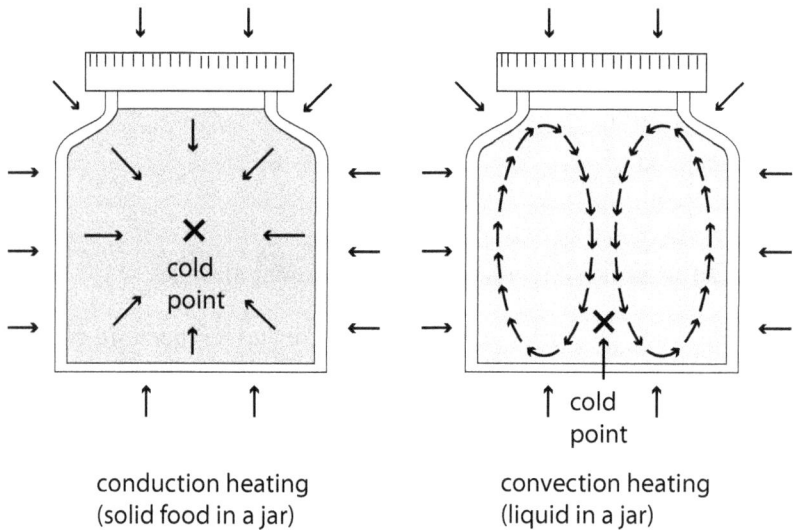

conduction heating
(solid food in a jar)

convection heating
(liquid in a jar)

Heat is transferred from the outside of the jar/can to the interior of the *solid* canned food through *conduction*. This is a slow method as the heat is transferred by molecule-to-molecule transfer. Meats, poultry, fish, potatoes and beets are heated by conduction. The last portion heated, the so called "cold spot" is usually the *geometric center* of the container.

Heat transfer in *liquids* is by *convection*. The convection method is *faster* as the heat is transferred by the moving currents of liquid itself. Meat broth or soup with a few solid pieces will heat much faster than a broth with solid chunks of meat. Therefore, where possible, it is important to have the food in smaller cuts and surrounded by liquid to allow these currents.

Fig. 5.11 Transfer of heat in solid food in a jar during steam processing.

Fig. 5.12 Transfer of heat in liquid or semi-liquid food in a jar during steam processing.

In liquid or semi-liquid foods the critical thermal point (*cold point*) is located about ⅓ of the height from the bottom of the container. A smoked sausage can be baked in a smokehouse to a safe internal meat temperature by raising the temperature to about 176-190° F, 80-88° C, which is a slow process. The same sausage immersed in hot water 176° F, 80° C will cook much faster as water conducts heat much faster than air. For that reason, the majority of processed meats are cooked in water. To take advantage of the shorter heating times of the convection method, liquid is added to solid chunks of food and a combination method (*conduction-convection*) is created. Liquid convection currents supply heat to solid food where the heat is transferred by a conduction method (molecule heat transfer). Heat penetration into meat and poultry is accomplished principally by *conduction*, with convection playing a lesser role. However, when products are packed in brine, broth or other liquids, convection heating plays a bigger role.

There are many cans on the market and they come in different shapes and sizes. A good understanding of the subject of heat transfer will make can selecting much easier.

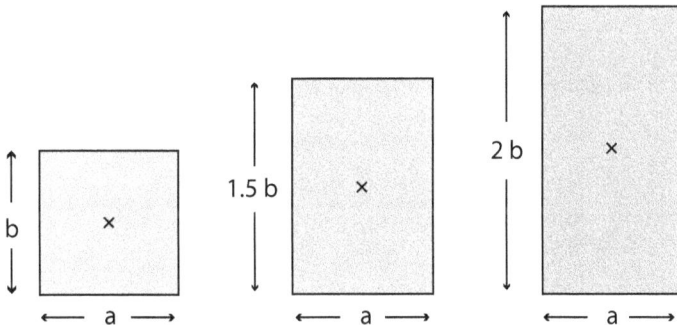

Fig. 5.13 Three cans with the same diameter, but different height. It takes longer for the heat to reach the "cold point" of the highest can than the lowest can.

Regulate heat to maintain a steady, constant pressure. Fluctuating pressure causes loss of liquid from glass jars and uneven cooking of contents.

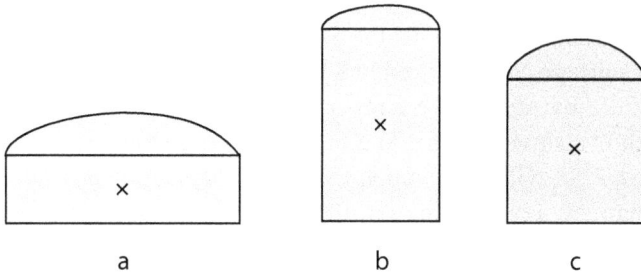

Fig. 5.14 The capacities of all cans are equal, however, the heat will penetrate faster to the cold point of flat cans "a" and "b" than to the square can "c."

Photo 5.4 It is quite obvious that heat transfer will be the fastest in the smallest jar.

Home and commercial canning equipment and methods differ greatly. Slower heating and cooling times with home equipment account for the greatest difference. Canning recipes for home canners are designed with safety in mind, taking into consideration that a hobbyist works without supervision from food inspectors.

The efficiency of heat transfer in containers depends on:

- The size and shape of the container - small diameter container will heat faster.

- Physical properties of the food product - food with a higher moisture content will heat faster. A freshly stuffed sausage will contain more moisture than a sausage that was conditioned for 2 hours at room temperature and then smoked for 3 hours.

- Heat transfer characteristics of the heating medium - water, broth, syrup, sugar, gravy, all those factors affect the heat transfer. Starchy ingredients absorb liquid during processing, and change the heat transfer process. When heated, the starch thickened gravy will change its state from a liquid to a more solid state and will affect convection currents and heating times. Fats and oils greatly retard thermal death of bacterial spores, vegetative cells and yeasts. For this reason fatty meat or fish requires longer processing times. Sodium chloride is the main factor controlling meat spoilage in regular meat processing. Elevated salt levels make production of dry hams or traditionally cured salami possible. In the canning process, however, salt is of little importance as heat is the main safety hurdle. When salt is applied up to 4%, the spores display increased resistance to heat. Only at levels over 8%, the resistance of spores to heat treatment decreases. Unfortunately, such high salt levels will make the product unpalatable.

- Heat transfer characteristics of the container - the thickness of the glass or a metal can, type of metal (steel, aluminum).

- The tightness of the pack will affect the movement of the currents as they will face less or more resistance. The way the material is cut will influence the movement of the current, for example vertically packed carrots versus diced carrots will experience heat transfer differently.

- Under processing can result in spoiled food, while over processing results in overcooked food.

4. Cooling.

There is a significant difference between commercial and home cooling processes. The thermally processed container should be cooled without delay to prevent the growth of *thermophillic* bacteria. As explained in the microbiology section, *thermophillic* bacteria like to grow at 122-150° F, 50-66° C. It is very likely that *thermophillic* bacterial spores will survive thermal treatment. When the temperature inside the can drops down to (122-150° F, 50-66° C), they will encounter favorable conditions to grow. For that reason commercial producers cool containers as fast as possible to about 95° F, 35° C. However, a rapid depressurization of the canner and introducing cold water induces a great strain on a container which still remains under high pressure. This leads to a great pressure differential between the canner and the inside of the container and results in the following problems:

- The contents of the *glass jar* will boil over through the still soft sealing compound. There will be less product inside, though the product will be fine and safe, as the sealant will reseal the lid when the pressure drops again. A stuck food particle, however, may prevent the seal.

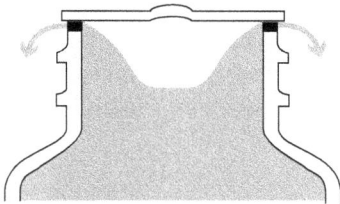

Fig. 5.15 Food boil over.

- Introducing cold water will shatter the hot jar due to thermal shock.
- The high pressure inside the *metal can* will strain the seams and the ends of the can may buckle. What is worse is the seam can be compromised and might not seal itself again, creating a safety risk. Smaller cans are usually not susceptible to this problem but any can from No. 3 (404 x 414) and bigger might end up with a damaged seal.

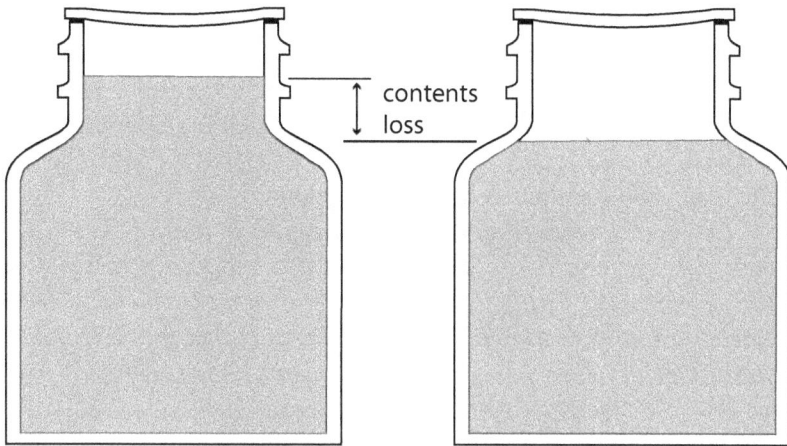

Fig. 5.16 Content loss due to boil over.

Commercial packers solve this problem by injecting compressed air first, and then introducing cold cooling water. The compressed air maintains the pressure inside of the canner at a level comparable with the pressure inside the container, and cold water lowers the temperature. This simple procedure requires large and expensive canners known as "retorts" and such equipment is beyond the reach of a hobbyist. However, pressure cooling protects jars from breaking and cans from buckling.

Cooling at Home

Glass Jars. Home pressure canners have no means for injecting compressed air and introducing cold water. There is no other practical solutions but to wait for the canner to depressurize itself, which usually takes about 30 minutes. Then the jars are taken out and left for cooling.

Metal Cans. It is possible to cool metal cans when using home pressure canners. Manufacturers specifically prohibit removing regulators from the steam vent pipe to drop the pressure rapidly in order to protect themselves from possible lawsuits. We all know the case when Stella Liebeck spilled hot coffee over herself at McDonalds, took the company to court and was awarded 3 million dollars. Since Liebeck, major vendors of coffee, have been subjected to similar lawsuits. Similar lawsuits against McDonald's in England failed.

Imagine manufacturers allowing people to remove steam pressure regulators from pressurized canners. There will be ten new lawsuits every day. Somehow, 70 years ago, the pressure canners were equipped with pet-cocks and home canners were gradually releasing pressure without injuring themselves. Well, we are a computerized society today, so we cannot handle simple manual tasks anymore.

Cooling procedure from the past: *If canning in tin cans - with No. 3 cans, let pressure return to zero, the same as for glass jars, before opening pet-cock. With smaller cans, the pet-cock can be opened gradually without waiting for the canner to cool and pressure return to zero. Open canner as soon as all steam has been released. Take cans out of canner and cool at once in clean, cold water, until the water turns luke warm.*

If the above procedure was performed and the cans were removed, their ends should be buckled due to the high pressure inside. This confirms the cans were sealed properly and are tight. As they cool down pressure will fall, a vacuum will form and the lid will become slightly concave.

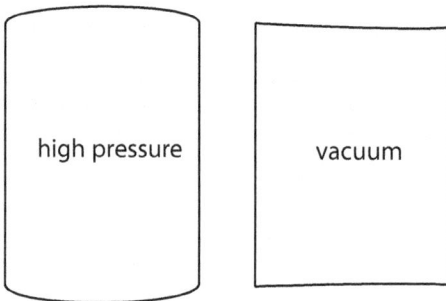

Fig. 5.17 Left-pressurized can, right-can with vacuum inside.

Summing it up: even though you could depressurize the canner faster, please follow the manufacturer's guidelines and wait until the pressure canner depressurizes to "0". However, once the canner is fully depressurized there is no harm in immersing metal cans in cold water to drop the temperature to 95° F, 35° C.

Cooling With Air

Cooling cans with an air-fan at room temperature is a waste of time as air is such a poor cooling medium. When the ambient temperature is low outside, for example in the winter, cooling cans in the air will be effective.

Summary on Cooling

Most likely after thermal process, the canner will be left alone to de-pressurize and then the jars will be removed and placed on a towel in an area without much draft. Then, they will be left overnight to cool, form a vacuum and create a strong seal. Metal cans can follow the same procedure, however, they can be moved around and placed in a drafty area. They are stronger and the seal is well formed by now.

There is a considerable advantage in a slow home canning process. The lack of cooling is offset by the additional processing time that the containers are submitted to. The long depressurizing step is nothing else as the additional heating step which starts at 240° F, 116° C and continues to about 212° F, 100° C. It corresponds to a temperature range of slow-barbecue and may be considered an extra cooking time. This adds considerable safety to a home canning process.

Storage

Canned products should be stored in a dry, cool place. Ideally, the storage area should have some air circulation, otherwise molds might grow on the outside of jars, screwbands, and metal cans might start to rust. Correct processing will kill spoilage bacteria and foods will not spoil, unless kept at temperatures over 95° F, 35° C. At these tempera-tures certain heat loving microorganisms, which could have survived the canning process, might start growing again. Nevertheless, certain chemical reactions which affect fat rancidity, food texture, color and nutritional value may occur when proper storage conditions are not met. The best storage temperature is about 35-59° F, 2-15° C, and the higher it is, the more unwelcome changes will occur in stored foods. As explained earlier, extremely high temperatures may create favor-able conditions for *thermophillic* bacteria that survived heat treatment to grow and spoil the food.

Freezing will affect texture quality due to water crystal formation inside of the food. This will make frozen food less palatable than properly stored canned food. Besides, this expansion of water may break glass jars and weaken the seals. Glass jars should be protected from light otherwise fats will develop rancidity and the quality of the product will suffer. At proper conditions meat and fish will keep well for 2-3 years. If jars must be kept in a very cold place, just insulate them by any practical means, for example, wrapping them in newspapers or keeping them in boxes covered with more newspapers or blankets.

Canning Meat in Glass Jars

    1. Preparation.

    2. Packing.

    3. Sealing.

    4. Thermal process.

    5. Cooling.

    6. Storing.

1. Read the manufacturer's instructions on the use of your pressure canner. Use only standard jars intended for home canning. These jars will have the manufacturer's name molded in the glass. Never use jars from commercial food products.

2. Either a raw pack or hot pack method may be used. Add salt (½ teaspoon per pint jar, 1 tsp for quart jar) and spices into the empty container. Leave 1 inch headspace in glass jars.

Raw pack - use raw meat, do not brown. Raw meat is packed directly into jars. Do not tightly pack the jars. Add only meat and seasonings, don't add liquid. Raw meat will release its own juices during the canning process.

Hot pack - brown meat before packing. Brown the meat in a skillet with a small amount of fat. The meat shrinks and can be packed more tightly. Save the liquid which will be added (hot) to the jars after packing. If more liquid is needed, add broth, tomato juice or water.

Note: it is common to sterilize jars when using a *water bath* canner. What about a pressure canner? Jars that will be processed in a pressure canner don't have to be sterilized as the high canning temperatures will sterilize them anyhow.

Run a plastic utensil around the inside of the jar to free bubbles and remove excess air.

3. Wipe off the rim of the jar and place the lid on top. Secure the lid with a screwband, finger tight.

4. Put 2 to 3 inches of water in the canner. The temperature of the water should be similar to the temperature of the product; this prevents stress on the glass jar. Place filled jars *on the rack*, using a jar lifter. Fasten the canner lid securely. Leave the weight off the vent pipe. Heat at the highest setting until steam flows freely from the vent pipe. This pushes the excess air out of the canner through the vent port. If the canner is not exhausted, the inside temperature may not correspond to the pressure on the gauge.

While maintaining the high heat setting, let the steam escape continuously for 10 minutes and then *place the weight on the vent pipe*. The canner will pressurize during the next 3 to 5 minutes. Turn the heat down slightly when the canner reaches about 7 PSI, then wait for the pressure to build up. *Wait until a dial indicates 11 PSI or the 10 pounds weighted gauge starts to jiggle.*

Now you can start timing the process.

- Process *meat and poultry* in half pints and pints for 75 minutes and quarts for 90 minutes.

- Process *fish* in half pint and pint jars for 100 minutes.

- Process *fish* in quart jars for 160 minutes.

Regulate heat under the canner to maintain a steady pressure at or slightly above the set gauge pressure. Quick and large pressure variations during processing may cause unnecessary liquid losses from jars. The pressure must be maintained in order to hold the lids in place during processing and in the early stages of cooling so that the pressure in the canner always exceeds the pressure inside the container. Follow the canner manufacturer's directions for operating your pressure canner.

Note: if at any time pressure goes below the recommended amount, bring the canner back to pressure and start the timing of the process over again, from the beginning (using the total original process time). This is important for the safety of the food.

5. When the timed process is completed, turn off the heat, remove the canner from heat if possible, and *let the canner depressurize. Do not force-cool the canner.* Allow the pressure to drop naturally, do not apply cold water, cold cloth or remove the weight until the pressure drops completely. Forced cooling may result in unsafe food if the cooling period has been included as a part of the thermal process. Cooling the canner with cold running water or opening the vent port before the canner is fully depressurized will cause loss of liquid from jars and seal failures. After the canner is depressurized and the pressure drops to 0, test the canner to ensure there is no more steam left. Slightly rock the weight, no steam should be released and no resistance should be felt. Remove the weight from the vent pipe or open the petcock. Wait 10 minutes, unfasten the lid, and remove it carefully. Lift the lid *away from you* so that the steam does not burn your face. Depending on the size of the canner it may take from 15 to 45 minutes for the pressure to drop to zero.

Remove jars with a jar lifter, and place them on a towel or a cooling rack. Do not retighten lids after processing, you may damage the seal. Let jars sit undisturbed to cool at room temperature for 12 hours. Never rush the cooling process or jars may break. Jars should be cooled in an area away from drafts. Air blowing on hot jars may also cause breakage. Jars will seal as they cool. The jar lid will pull in when a vacuum is formed. Test the seals.

6. Label the jars with the type of meat, date, processing time and pounds of pressure. Store in a cool dark place. The screwbands may be removed. *Do not freeze* jars as this may compromise the seal.

Headspace Gauge

Photo 5.5 When the clear plastic teeth of the headspace gauge contact the liquid level in a container, the diffused light indicates the headspace to 1/16 inch.

Canning Meat in Cans

Whereas canning in jars can be accomplished by using only three sizes of a glass jar: quart, pint and ½ pint; canning in tin cans can be confusing due to dozens of can styles and sizes. For these reasons we are going to advocate only a few can sizes. The selection is based on USDA research papers which recommended No. 2, No. 2.5 and No. 3 cans as suitable for home canning. We also include smaller cans which are popular among home canners in Pacific West states, Canada and Alaska for canning fish. Furthermore, we have chosen can sizes with as few different diameters as possible, so that they can be sealed with just one sealer. When more than one can diameter is used, we have made sure that the sealer can seal them by installing a different chuck, a very simple operation to perform.

Low acid foods such as meat, poultry and fish must be processed in pressure canners.

1. Preparation.
2. Packing.
3. Exhausting cans.
4. Sealing.
5. Thermal process.
6. Cooling.
7. Storing.

It is important to note that *the seal is formed differently in cans than in jars*.

In glass jars the air is forcefully expelled during the heating process by the mounting pressure inside. The lid is on top of the jar and although it is secured with a ring, the seal has not formed permanently yet. During heating the soft seal allows pressurized air to escape from the jar. Upon cooling *the vacuum inside the jar sucks in the lid and the softened seal makes a tight closure upon cooling*.

In cans the sealing machine forms a tight connection between the lid and the body of the can. During heating the expanding air will have no way to leave and will create a lot of pressure on ends and seams of the can. In addition, the air discolors food and slows down heat penetration. So the first step in the canning process is to exhaust as much air as possible from the can before sealing.

1. Preparation - rinse the cans, but do not rinse the lids. The sealing compound is very delicate and washing lids may compromise sealing the can. Cut the meat into 1 inch cubes or any convenient size for your can. *Precook meat* by roasting, stewing or frying in oil or fat.

2. Packing - pack meat *loosely* into cans. Leave a ¼ inch headspace between the meat and the top of the can. Salt, pepper and spices are added on top of the packed meat. Fill each packed can to ¼ *inch from top* with boiling meat juices, broth, water or tomato juice. It is generally accepted that headspace should not exceed ten percent of the total can volume. Remove air bubbles by running a plastic knife around the inside of the can, to release bubbles and any trapped air.

3. Exhausting.

It is important that as much air as possible is removed from the cans. Syrups and brine should be added as hot as possible.

3.1 Place the canner rack on the bottom of the canner. Exhaust only one layer of cans at a time. Place *open*, filled cans inside the water in the canner. Add enough water to come *half way* up the sides of the *open* cans filled with meat.

3.2 If you have more than will fit in one layer, put the second layer in a *different pan and heat it separately*. Place cans in the water where the liquid comes about halfway up the sides. Set the open cans filled with meat in a single layer in simmering water in open roasting pans on the stove. Do not cover the canner or the pan, otherwise condensing moisture will drop into the cans.

3.3 Adjust the temperature so the water comes to a *gentle boil*. Check the temperature of the meat in the cans with a meat thermometer. The internal temperature of the meat in the center of the can must reach 170° F, 77° C.

3.4 Remove cans from the boiling water using a jar lifter. Carefully clean the edges of each can with a towel.

4. Sealing - center each hot can of meat on the can sealer and place a lid on top. Bring the top of the can sealer down until it is completely engaged. Always refer to the instructions that came with your can sealer. Inspect each can to ensure that it is completely sealed. If a seam is formed incorrectly, the meat must be removed from the can, packed into a new can, exhausted to 170° F, 77° C and resealed.

Correctly sealed cans need to be kept hot. They may be placed in hot water in the canner.

5. Thermal Process.

5.1 Add 2-3 inches of water to the canner and place the rack. The water temperature should be similar to the temperature of the product in the can. The procedure for canning metal cans is similar to the one for canning glass jars. Cans may be stacked in the canner. When your canner is filled, fasten the lid securely. Turn on the heat.

5.2 Leave the weight off the vent port or open the pet-cock. Heat at the highest setting until steam flows freely from the open pet-cock or vent port. This pushes the excess air out of the canner through the vent port.

While maintaining the high heat setting, let the steam flow (exhaust) continuously for 10 minutes and then place the weight on the vent port or close the pet-cock. The canner will pressurize during the next 3 to 5 minutes. *Wait until a dial indicates 11 psi or the 10 pounds weighted gauge starts to jiggle.*

5.3 Now, you can start timing the process. Write down the starting and ending time.

Cans and lids: Tall, 1 pound (size: 301 x 408) or flat, ½ pound (size: 307 x 200.25), also called Alaska salmon cans, are used for canning meats.

According to the University of Alaska in Fairbanks, plan to use 1 to ½ pounds of trimmed meat per 1-pound can; ¾ pound of trimmed meat per ½ pound can.

Process one pound cans for 90 minutes at:

10 psi - weighted gauge
11 psi - dial gauge

Process half pound cans for 70 minutes at:

10 psi - weighted gauge
11 psi - dial gauge

6. Cooling - turn off the heat, remove the canner from heat if possible, and *let the canner depressurize. Do not force-cool the canner.* After the canner is depressurized and the pressure drops to 0, test the canner to ensure there is no more steam left. Slightly rock the weight, no steam should be released and no resistance should be felt.

Remove the weight from the vent port or open the pet-cock. Depending on the size of the canner it may take from 25 to 60 minutes for the pressure to drop to 0 PSI. Wait 10 minutes, unfasten the lid and remove it carefully. Lift the lid away from you so that the steam does not burn your face. After the pressure is reduced to zero, the cover of the pressure cooker may be removed and *the cans water or air-cooled*. Jars must be air-cooled, as a sudden temperature drop may cause the glass to crack.

Cooling Glass Jars in Home Production

Jars are at a disadvantage as they face a thermal shock. Placing a hot jar in cold water will crack the glass. The glass jars are usually left undisturbed in the air for 12 hours *to cool by themselves*. However, there is a safety benefit in allowing the glass jars to cool by themselves; this fact is well documented in laboratory research studies.

From 930 Bulletin Home Canning Processes for Low Acid Foods:

"These tests show that when foods are home canned in glass jars the long cooling periods required contribute significantly to the lethal value of processes. With vegetables in pint jars the sterilizing value of the cooling period averaged 50% of the total. In quart jars an average of 36 percent of the process value was contributed by the cooling period. In No. 2 and No. 2.5 tin cans the corresponding averages were 15 and 11 percent, respectively.

Relatively high values of cooling periods for glass packs did not, however, lead to generally shorter processes as compared with packs in tins. The exhaust given the latter before sealing resulted in higher initial temperatures which tended to balance the sterilizing value of the cooling period for processes in glass containers."

Cooling Cans in Home Production

Air-cooling facilitates the transfer of heat into the air. The cans are piled in rows, allowing sufficient space between rows for efficient air circulation. Air fans will greatly improve heat transfer. This type of cooling is well suited to products in small cans, in other words, home production.

At the end of the canning process the product's temperature is 240° F, 116° C. Depending on the size of the canner it may take 15-45 minutes for the canner to drop to 0 PSI and 212° F, 100° C. This temperature is equivalent to barbecuing at a low setting and this additional time still cooks the product. *After depressurizing*, the hot *cans* may be taken out of the canner and allowed to cool by themselves or be inserted in cold water. They already have a strong seal and can be moved around. Smaller cans, No. 2, No. 2.5 and even No. 3 can be immersed in cold water and cooled to 95-105° F, 35-40° C, but not lower. Then, they must be placed on a rack and *allowed to dry*. Let cans sit undisturbed to cool at room temperature until dry. This is the same procedure as used for glass jars, however, they have a strong seal and can be moved around. You may follow the same procedure for glass jars and let cans cool by themselves for 12 hours. You gain additional safety by cooking the product longer.

7. Check the seams of all cans. Seams should be flat and smooth, with no pointed or rough edges. No leakage should be seen around the can edges. If some cans did not seal there are two options:

- Freeze the contents of the unsealed can, or refrigerate the product and use within 3 to 4 days. Alternatively, the meat may be reprocessed in a new can within 24 hours of the initial processing period.

- Reprocess the meat in a new can within 24 hours of the initial processing period. Reprocessing does not affect the safety of the meat, however, it will affect its texture, after all, the product is cooked a second time.

Do not freeze cans as this may compromise the seal. Label the cans with the type of meat, date, processing time and pounds of pressure. You can use a permanent marker to label cans. Store in a cool dark place.

The cooling step is performed differently by commercial producers who use different pressure canners called retorts or autoclaves. *Water cooling* is used because water transfers heat much faster than air. Rapid cooling of the container contents prevents softening in texture or change in color of the food. Commercial producers are well aware of the importance of cooling products rapidly in order to ensure the highest quality of canned foods.

The cans are loaded onto metal baskets and are inserted into retorts which can be of the horizontal or vertical type. High temperatures in those units are generated by steam or hot water which are delivered in pipes. When the sterilization process ends, cold water is introduced into the retort. Only potable quality water may be used for cooling. Commercial producers use chlorinated water to eliminate the risk of bacterial contamination in case the seal may develop a leak.

The cold water causes the hot steam to condense *creating an immediate pressure and temperature drop* inside. The cans should be rapidly cooled to 95-105° F, 35-40° C but not much lower. This is to ensure that some residual heat still remains to dry the can exterior. Only then dry cans can be cased, stored or shipped as the possibility of post-process contamination in dry cans is remote.

Very rapid cooling, however, may create a problem, because the pressure inside cans is still high and so is the temperature. There is a significant *pressure difference* between the inside of the container and the inside of the retort. This may weaken the seal or even deform the cans, especially when they are of a large size. In a tall retort, the height of cold water supplies higher pressure and will counteract the pressure inside the cans. In modern equipment the "pressure-cooling" method is practiced; a blast of compressed air is delivered into the retort when the cooling starts. This maintains pressure in the retort and counter-balances the pressure within the can, however, the temperature drops down due to the introduction of cold water.

Storage

Store cans at room temperature. What we want to mention is that in many cases, for example in tropical countries, canned products might be kept at temperatures higher than 77° F, 25° C. Although bacteri-al spores will be eliminated during proper processing times at 240° F, 116° C, there are certain bacteria (*thermophillic bacillus* strains) which are extremely heat resistant and they will survive. They nor-mally don't grow below 77° F, 25° C, even if present in a sealed con-tainer. However, when storage temperatures will reach 95° F, 35° C or higher, those bacteria will grow and given time will spoil the food. To eliminate this possibility, the processing times and temperatures must be increased even more. It should be noted that those bacteria spoil the food only and don't pose grave risks to our safety.

Ensuring High-Quality Canned Foods

Canning is a method of food preservation and its primary aim is to prevent the growth of microorganisms that would spoil the food and create danger to consumers. Canning does not create super quality food, it simply preserves the food that we make. If we prepare a good dish, canning will definitely preserve its quality for a long time. If we simply throw in some meat without spices, salt or broth, we would preserve it, but it would hardly be a culinary masterpiece. In other words, all good cooking principles should be applied not only for the safety of the product, but also for its taste and flavor.

Begin with good-quality fresh foods suitable for canning. Quality varies among varieties of fruits and vegetables. Discard diseased and moldy food. Trim small diseased lesions or spots from food. Can fruits and vegetables picked from your garden or purchased from nearby farmers when the products are at their peak of quality-within 6 to 12 hours after harvest for most vegetables. For best quality, apricots, nectarines, peaches, pears, and plums should be ripened one or more days between harvest and canning. If you must delay the canning of fresh produce, keep in a shady, cool place. Many fresh foods contain from 10 percent to more than 30 percent air. How long canned food retains high quality depends on how much air is removed from food before jars are sealed.

Maintaining Color and Flavor in Canned Food

To maintain good natural color and flavor in stored canned food, you must:

- Remove oxygen from food tissues and jars.
- Quickly destroy the food enzymes.
- Obtain high jar vacuums and airtight jar seals.

Follow these guidelines to ensure that your canned foods retain optimum colors and flavors during processing and storage:

- Use only high-quality foods which are at the proper maturity and are free of diseases and bruises.
- Use the hot-pack method, especially with acidic foods to be processed in boiling water.
- Don't unnecessarily expose prepared foods to air. Can them as soon as possible.

- While preparing a canner load of jars, keep peeled, halved, quartered, sliced, or diced apples, apricots, nectarines, peaches, and pears in a solution of 3 grams (3,000 milligrams) ascorbic acid (vitamin C) to 1 gallon of cold water. This procedure is also useful in maintaining the natural color of mushrooms and potatoes, and for preventing stem-end discoloration in cherries and grapes.

- Fill hot foods into jars and adjust headspace as specified in recipes.

- Tighten screwbands securely.

- Process and cool jars.

- Store the jars in a relatively cool, dark place, preferably between 50-70° F (10-21° C).

- Can no more food than you will use within a year.

You can get ascorbic acid in several forms:

- Pure powdered form - seasonally available among canners' supplies in supermarkets. One level teaspoon of pure powder weighs about 3 grams. Use 1 teaspoon per gallon of water as a treatment solution.

- Vitamin C tablets - economical and available year round in many stores. Buy 500 - milligram tablets; crush and dissolve six tablets per gallon of water as a treatment solution.

- Commercially prepared mixes of ascorbic and citric acid are seasonally available among canners' supplies in supermarkets. Sometimes citric acid powder is sold in supermarkets, but it is less effective in controlling discoloration. If you choose to use these products, follow the manufacturer's directions.

Equipment and Methods not Recommended

Open kettle canning and the processing of jars in conventional ovens, microwave ovens, and dishwashers are not recommended, because these practices do not prevent all risks of spoilage. Steam canners are not recommended because processing times for use with current models have not been adequately researched. Because steam canners do not heat foods in the same manner as boiling-water canners, their use with boiling-water process times may result in spoilage. It is not recommended that pressure processes in excess of 15 PSI be applied when using new pressure canning equipment. So-called canning powders are useless as preservatives and do not replace the need for proper heat processing. Jars with wire bails and glass caps make attractive antiques or storage containers for dry food ingredients but are not recommended for use in canning. One-piece zinc porcelain-lined caps are also no longer recommended. Both glass and zinc caps use flat rubber rings for sealing jars, but too often fail to seal properly.

Low-Acid Acidified Foods

A considerable safety margin can be introduced into processing time by acidifying the product. If the low acid food, for example cucumber, is acidified to a pH of 4.6 or less, it crosses the threshold that separates low acid foods from the high acid foods. In other words it becomes a high acid food and as such, it requires less severe thermal treatment to achieve sterility. It can, theoretically, be sterilized in a water bath canner (212° F, 100° C). As the thermal resistance of bacterial spores decreases in an acidic environment, they should not grow in foods with a pH below 4.6. Thermal treatment is needed to kill only spoilage bacteria, molds, yeasts and enzymes, all of which can be killed at 212° F, 100° C.

Adding citric acid, lactic acid, lemon juice or vinegar will lower the pH of the product and cooking media. Chicken that was marinated overnight with salt, vinegar, white wine or lemon juice will acquire some acidity and will be more hostile to any bacterial spores than a fresh chicken. A pH meter is needed to measure acidity levels accurately, the pH color strips are suitable for checking water pH in a fish tank or for general less critical applications.

Meats, poultry, fish, vegetables and dairy products fall into a pH range of 5.0-6.8. These are low-acid foods so they must be processed at 240-250° F, 116-121° C, unless they become acidified to such an extent that the pH equilibrium of the finished product is pH 4.6 or lower. Acidified foods do not automatically fall into the high-acid food category, they become *low-acid acidified* products.

As mentioned earlier, acidifying foods and establishing new processing times must be left to properly trained persons. A home canner should follow the rules established by USDA guidelines without regard to the extra acidity that he may have introduced.

Vegetables

Vegetables are low-acid foods and are subject to the same regulations as meat, poultry and fish. As vegetables grow in soil they usually contain large numbers of microorganisms. As explained earlier, the soil is the major carrier of *Cl.botulinum* spores. Vegetables are always washed but that removes only a part of the microorganisms and soil. The next step, known as *blanching* removes, kills or inactivates the majority of microorganisms present in vegetables. However, the ones that pose the highest security risk, such as *Cl.botulinum* survive as bacterial spores. Vegetables must be exposed to 240-250° F, 116-121° C heat treatment like all other low-acid foods in order to kill *Cl.botulinum* spores. *Fermented* vegetables such as sauerkraut and pickles can be pasteurized at lower temperatures as specified in the USDA guides.

High temperature thermal process often adversely affects delicate textured food like vegetables so commercial producers acidify the product to below pH 4.6. That allows them to process vegetables at pasteurization temperature. Look at the products in a supermarket; most food, for example, different types of pickles are *acidified* with vinegar in order to process them at lower temperature. For example, pH of cucumbers varies from 5.1-5.7, but pH of canned dill pickles is 3.2-3.5. Low-acid products acidified recipes are designed by the *processing authority* and approved by the FDA. However, the United Stated Department of Agriculture Bulletin 539 lists many vegetable recipes that may be safely canned at home. All information in this book applies to both vegetables and meat products and that includes equipment, processing steps and testing methods.

The table below lists some common fruits and vegetables according to their pH. Acidity of foods depends upon many factors such as variety, maturity and growing conditions of the product. For these reasons, the pH of food is usually within a range of values.

Vegetables	pH	Vegetables	pH
Artichokes	5.6	Spinach	5.5 - 7.2
Asparagus	4 - 6	Tomatoes, whole	4.2 - 4.9
Beans	5.7 - 6.2	Turnips	5.2 - 5.5
Beets	4.9 - 5.6	Zucchini, cooked	5.8 - 6.1
Brussel sprouts	6.0 - 6.3	**Fruits**	**pH**
Cabbage	5.2 - 6.0	Apples	3.3 - 3.9
Carrots	4.9 - 5.2	Apricots	3.3 - 4.0
Cauliflower	5.6	Bananas	4.5 - 5.2
Celery	5.7 - 6.0	Cantaloupe	6.17 - 7.13
Chives	5.2 - 6.1	Dates	6.3 - 6.6
Corn	6.0 - 7.5	Figs	4.6
Cucumbers	5.1 - 5.7	Grapefruit	3.0 - 3.3
Eggplant	4.5 - 5.3	Lemons	2.2 - 2.4
Horseradish	5.35	Mangos	3.9 - 4.6
Kale, cooked	6.4 - 6.8	Melons	5.5 - 6.7
Leeks	5.5 - 6.0	Nectarines	3.9
Lettuce	5.8 - 6.0	Oranges	3.1 - 4.3
Lentils, cooked	6.3 - 6.8	Papaya	5.2 - 5.7
Mushrooms, cooked	6.2	Peaches	3.4 - 3.6
Okra, cooked	5.5 - 6.4	Persimmons	5.4 - 5.8
Olives, green	3.6 - 3.8	Pineapple	3.3 - 5.2
Olives, ripe	6.0 - 6.5	Plums	2.8 - 4.6
Onions	5.3 - 5.8	Prunes	3.1 - 5.4
Parsley	5.7 - 6.0	Tangerines	4.0
Parsnip	5.3	Watermelon	5.2 - 5.8
Peas	5.8 - 7.0	**Berries**	**pH**
Pepper	5.15	Blackberries	3.2 - 4.5
Pimiento	4.6 - 4.9	Blueberries	3.7
Potatoes	6.1	Cherries	3.2 - 4.1
Pumpkin	4.8-5.2	Cranberries	2.4
Radishes, red	5.8 - 6.5	Currants, red	2.9
Radishes, white	5.5 - 5.7	Gooseberries	2.8 - 3.1
Rhubarb	3.1 - 3.4	Grapes	3.4 - 4.5
Rice, cooked	6.0 - 6.7	Raspberries	3.2 - 3.7
Sauerkraut	3.4 - 3.6	Strawberries	3.0 - 3.5

Data from the FDA/CFSAN - Bad Bug Book - pH Values of Various Foods.

Photo 5.6 Well stocked kitchen pantry.

Chapter 6

Preparing Food For Canning, General Guidelines

Meat and Poultry

Keep all meat at low temperature until ready for processing. If meat must be kept for longer than a few days, freeze it. Trim meat off gristle and fat. Fat left on meat will melt and climb the sides of the jar during canning. If it comes in contact with the sealing edge of the lid, the jar may not seal. How you cut your meat will affect its use.

If possible, always use fresh meat as it has the lowest bacteria count. Fresh home-slaughtered red meats and poultry *should be chilled and canned without delay.* Keeping meat in a refrigerator significantly slows down the growth of bacteria, but they still manage to multiply. The more bacteria in the meat, the longer time is needed to eliminate them, even at higher temperatures. Frozen meat may be canned, but it does not make a high quality product. For best results it is better to cut frozen meat into strips 1 to 2 inches thick and plunge into boiling water. Simmer until the color of the meat has almost disappeared, then immediately pack and process.

When you must use the frozen meat thaw it in the refrigerator or place the wrapped meat under cold, running water. Trim away all freezer burn. Freezer burn occurs when frozen food has been damaged by dehydration and oxidation, due to air reaching the food. It is generally induced by substandard (non-airtight) packaging. Freezer burn does not affect the quality of the final product, as long as it is removed prior to processing. Smaller diameter cuts may be thawed in a micro-wave. Thawing results in a loss of natural meat juices and results in about 1-3% weight loss as some of the internal water leaks out.

Meat Color

The color of *fresh meat* is determined largely by the amount of *myoglobin* a particular animal carries. The more *myoglobin* the darker the meat, it is that simple. Going from most to least, the range of myoglobin containing meats are listed in descending order: horse, beef, lamb, veal, pork, dark poultry and light poultry. The amount of myoglobin present in meat increases with the age of the animal. Different parts of the same animal, take the turkey for example, will display a different color of meat. Muscles that are exercised frequently such as legs need more oxygen. *As a result they develop a darker color* unlike the breast which is white due to little exercise. This color is pretty much fixed and there is not much we can do about it unless we mix different meats together.

The color of *cooked* (uncured) meat varies from greyish brown for beef and grey-white for pork and is due to denaturation (cooking) of *myoglobin*. The red color usually disappears in poultry at 152° F (67° C), in pork at 158° F (70° C) and in beef at 167° F (75° C). The color of *cured* meat is pink and is due to the reaction between nitrite and *myoglobin*. The color can vary from light pink to light red and depends on the amount of *myoglobin* a particular meat cut contains and the amount of nitrite added to the cure.

Nitrates

Adding nitrates to meat offers many benefits:

- Meat becomes pink.
- Acquires a slightly different flavor.
- Slows down oxidation.
- Inhibits *Cl.botulinum* from growing.

Sodium nitrite, commonly used as cure #1, is the strongest agent for preventing growth of *Cl.botulinum* spores during smoking meat. It is also added to impart the pink color to processed meats like ham and sausages. You wouldn't like to eat gray ham, would you? Well, without sodium nitrite a roasted leg of pork is just a roasted leg of pork, once the sodium nitrite is added it becomes pink ham with its distinctive flavor that sodium nitrite also provides. Adding sodium nitrite (cure #1) to canned meat will definitely lower the resistance of bacterial spores to thermal processing, but as explained earlier a hobbyist should still process foods according to the USDA guidelines.

Nitrate Safety Concerns

There has been much concern over the consumption of nitrates by the general public. In the 1970's much research was done on the effects of nitrates on our health. Millions of dollars were spent, many researchers had spent long sleepless nights seeking fame and glory, but no evidence was found that when nitrates are used within the established limits they can pose any danger to our health.

A review of all scientific literature on nitrite by the National Research Council of the National Academy of Sciences indicates that nitrites does not directly harm us in any way. All this talk about the danger of nitrite in our meats pales in comparison with the amounts of nitrates that are found in vegetables that we consume every day. The nitrates get to them from the fertilizers which are used in agriculture. Don't blame sausages for the Nitrates you consume, blame the farmer. It is more dangerous to one's health to eat vegetables on a regular basis than a sausage.

You do not need to use nitrites, however, there are many cases when you want meat to stand out, you want it to be the show piece. The example is the red color of tongue in blood sausage or red meat in a headcheese. If you want your canned meat be of red color, use sodium nitrite (cure #1), it is that simple.

Curing is an important part of sausage making technology, so we will touch upon it very briefly as this book is about canning meats. If you want to study curing in more detail, read our 720 page book "Home Production of Quality Meats and Sausages."

What is Curing?

In its simplest form the word 'curing' means 'saving' or 'preserving' and the definition covers preservation processes drying, salting and smoking. When applied to home made meat products, the term 'curing' usually means *'preserved with salt and nitrite.'* When this term is applied to products made commercially it will mean that meats are prepared with salt, nitrite, ascorbates, erythorbates and dozens more chemicals that are pumped into the meat.

There are dry and wet methods (brine) of curing. Meat for sausages is usually diced into 1-2 inch pieces, mixed with salt and cure #1 (2.5 g = ½ tsp of cure #1 per 1 kg of meat) and left for 72 hours. Larger cuts of meat will benefit from injecting them with curing solution at 10% of solution per weight of the fresh meat.

Mild curing solution (21 salinometer degree):
¾ cup of salt, 1 gallon of water, 2 tsp. of cure #1.

Inject meat or chicken with 10% of curing solution in relation to the weight of the meat, eg. 100 ml brine for 1 kg (2.2 lb) chicken. Immerse chicken in remaining brine and cure overnight in refrigerator.

There may be circumstances when there is no time to cure meat properly. In such cases, the curing process, especially the development of color will benefit greatly from cure accelerators, the simplest one being ascorbic acid which is vitamin C. When added to finely ground or emulsified products (e.g., luncheon meat), they can be canned almost immediately, and a uniform color will be attained. A vitamin C tablet may be pulverized and applied to meat. It is usually applied at 0.1%, e.g., 1000 mg vitamin C per 1 kg of meat.

Injecting meat with curing solution will increase its juiciness by decreasing cooking loss during precooking. It will also facilitate thermal processing as more meat will be subjected to convection heat transfer. Lastly, sodium nitrite is the strongest agent that inhibits the growth of *Cl.botulinum* spores and that is why it is always added to meats that would be smoked. Depending how the smokehouse is designed, there could be a little air inside or none at all creating favorable conditions for bacteria spores to grow.

Precooking Meat

Precooking meat prevents meat pieces from sticking together. It also provides better appearance and flavor. Precooking results in size reduction allowing meat cuts to be packed more tightly. Meats can be packed directly into containers, packed with liquid or mixed with vegetables, for example a stew. Ground beef can be mixed with spices (chili con carne) or mixing with spices and beans (chili con carne with beef). Some products such as pates or luncheon meat require additional preparation steps.

Liver Spreads

Liver spreads are nothing else than liver sausages which are not stuffed in casings. The most famous is the true French pate known as foie gras. Foie gras is made from the liver of forcefully fed geese. Such a liver is about three times bigger than usual and carries a distinctive flavor and creamy texture. Of course livers other than goose can be used for making quality liver spreads which may also be called pastes or pates. The technology for making canned liver pates is not different from that of fresh pates. Liver pates can be divided into two categories: fine and course. The degree of comminution, or in other words the particle size, will classify the product. The finest pates require emulsifying meat in a bowl cutter which is then followed in a colloidal mill. Those machines are cost prohibitive to be used at home, however, a great pate can be produced by grinding meat through a small plate and then emulsifying in a kitchen food processor. The technology for making liver products is described in *"Home Production of Quality Meats and Sausages"* and can also be accessed online on our website: *http://www.meatsandsausages.com/sausage-types/liver-sausage*

Luncheon Meats

The technology of making luncheon meats follows basic sausage making steps: meat selection, grinding, mixing and filling into containers. Depending on particle size the final product can have a consistency of a thick paste or a fine pate. Grinding through ⅛ inch plate will make it a paste, following in a food processor will make it a fine pate. If only the grinder is available, grind it the first time through a small plate, place in a freezer for 30 minutes, then grind it again.

Note: meats with much connective tissue, such as tendons, sinews, and skins contain collagen which is a strong binder. Upon heating, the collagen forms a gelatin which contributes to a better mouthfeel and texture of the product. Liver sausages, meat jellies or luncheon meats will benefit greatly from getting a part of collagen rich meat. Such meat known as grade III is ground with a small plate (⅛", 3 mm) or emulsified. Recipes call often for pork skins. This is not for quality reasons, but for gelling properties of pork skin. You can replace half of pork skins with meat from pork or beef legs, soft beef connecting tissues (sinews, tendons) or pork ears without bone tissue.

Poultry and Game Birds

Poultry meat may be canned with or without bones. Meat of wild birds is darker as it contains more myoglobin. This is directly related to the plenty of exercise the birds are subjected to. Some game birds exhibit strong flavor, especially water fowl, but that can be corrected by soaking birds for 1 hour in a brine made from 1 tablespoon of salt and 1 quart of water. Then they should be rinsed. If the birds were soaked in brine, don't add salt when filling containers.

Fish

When you catch fish, handle them with care to avoid bruising. Beware that exposure to the sun or heat may cause the quality of the fish to deteriorate. Bleed fish immediately after catching to increase the storage life. Remove internal organs and rinse the fish inside and out. Keep fish iced, refrigerated, or frozen until ready to process. Keep fish between 32-40° F (0-4° C) for no longer than one to two days.

You can use either fresh fish or frozen fish for pressure canning. Many Alaskans freeze their catch for up to one year. When fishing season arrives again the fish remaining in the freezer are canned. This gives the fish an effective shelf life of two years. When using frozen fish thaw it in the refrigerator or place the wrapped fish under cold, running water.

There is little we can do to control the composition of the fish flesh as in most cases fish are harvested wild. The living fish flesh is bacteria free. Bacteria is present on the skin, gills, and in the viscera, however, they cannot penetrate the flesh while the fish is alive. The fish flesh becomes contaminated with bacteria during handling and preparation. The sooner the fish can be processed the smaller the number of microorganisms it will contain. Poor quality fish that has been invaded by microorganisms can be characterized by the presence of slime, discoloration of the gills and eyes and loss of flesh texture.

To minimize microbial contamination, the fish surface must be scrupulously washed, gutted and rinsed inside. The gills must always be removed. The fish that will be canned must have the scales removed. The scales will fly everywhere so it is wiser to perform this operation outside. After cleaning, the fish has to be washed again. Previously frozen fish can be thawed, then brined and smoked.

Lobster and crab spoil very quickly after death. For this reason lobsters and some crabs are cooked alive in boiling water or weak brine (3-5%) to inactivate enzyme activity.

Shrimp can be cooked and peeled, or peeled raw and then cooked to harden the texture and cause the shrimp to curl.

Oysters and clams often contain mud or sand on the exterior and should be well washed before opened. Precooking in boiling water kills the animal, opens the shell and firms the meat.

There are circumstances in which a canner will select a process which is more severe than that required for commercial sterility, as for instance occurs when bone softening is required with salmon or mackerel.

Large fish are precooked whole in a steam kettle or cut into sections and precooked in brine.

Small fish are precooked in steam which removes moisture and oil that otherwise will be released in the container during thermal processing. That would adversely affect the texture, appearance and flavor of the product. This cooked out liquid should be drained away.

Precooked fish should be cooled as fast as possible to firm up the flesh to prevent breaking up the flesh during packing into the containers. The packing should follow immediately after.

Brining

Fish will benefit from immersing them in a strong brine (80 salinometer degrees), even for a short time. This toughens the surface of the flesh and removes traces of blood. The whole fish may be brined for 45 minutes, small fish or fillets for 5-15 minutes.

Smoking

Fish like other meats can be smoked by different smoking methods. Smoke temperature and the length of smoking will influence the taste of the fish. All fish may be smoked, but the fatty ones absorb smoke better, stay moister during smoking and taste better.

Ingredients

Salt, spices and condiments will improve flavor. Oil, brine, tomato juice, water or sauces improve appearance and flavor and enhance the heat penetration during thermal processing.

Packing Fish

Commercially canned fish is packed tightly to prevent it from shaking around and breaking into small pieces. It is also weighed to meet the requirements on the label. Fish that is processed at home must be cut in a way to best utilize the height of the container. Fillets can be rolled around the can, the shorter ones going inside the can. That is why the 307 x 22.25 can is so well adapted for canning fish.

The tall salmon can, 301 x 408, is another can that is great for canning a variety of fish. Both of those 2-piece cans are tapered which makes them especially attractive for storing at home as they nest inside each other.

Fish packed in jars should be neatly arranged to take advantage of the display properties of the glass, however, due to the shape of a glass jar it is harder to neatly remove the delicate product.

Vegetables

Vegetables must be cleaned by soaking in water, spray washing or both. Then they are peeled, trimmed, and the large ones are reduced in size to make them suitable for packing. Next comes blanching.

Blanching is a cooking process wherein the food substance, usually a vegetable or fruit, is plunged into hot water (close but below boiling point) for a few minutes only and plunged into iced water or placed under cold running water to stop the cooking process.

Blanching liberates air and other gases which are present in cells of fruits and vegetables. If they are not removed prior to sealing, the air trapped in vegetables will be released into the headspace and will affect heat transfer, cause product oxidation and even internal corrosion of the containers. Blanching reduces the amounts of microorganisms and pesticides, improves the washing process and softens the skin for peeling. It also deactivates some of the enzymes that might affect the flavor, color and texture of the product.

Chapter 7

Recipes

There is a small dilemma when the subject of low-acid recipes comes around as there are very few recipes available for a home canner. We have written eight books on processing meats and sausages which together contain almost one thousand recipes, yet not being a *processing authority*, we do not write recipes for low-acid canned products. We will, however, quote a low-acid recipe that comes from a *reputable source,* like the USDA's canning guides and bulletins.

Commercial packers have recipes that are either designed by the *processing authority* or the commercial plant may have a *processing authority* with a properly equipped lab to take care of the scheduled processes at moment's notice. Needless to say, such recipes are considered the trade secrets.

CFR 21, §118.83 A *processing authority* is a qualified person having expert knowledge of thermal processing requirements for low-acid foods in hermetically sealed containers and *having adequate facilities for making such determinations.*

Our note: each new product (recipe) must be designed by the *processing authority* and the canner must submit the scheduled process to FDA (FDA Form 2541 a) for the approval before the product can be distributed. The scheduled process is established for each low-acid food in each container size.

§108.35, 2 (ii) If a packer intentionally makes a change in a previously filed scheduled process by reducing the initial temperature or retort temperature, reducing the time of processing, or changing the product formulation, the container, or any other condition basic to the adequacy of scheduled process, he shall prior to using such changed process obtain substantiation by qualified scientific authority as to its adequacy. Such substantiation may be obtained by telephone, telegram, or other media, but must be promptly recorded, verified in writing by the authority, and contained in the packer's files for review by the Food and Drug Administration.

Any intentional change of a previously filed scheduled process or modification thereof in which the change consists solely of a *higher* initial temperature, a *higher* retort temperature, or a *longer* processing time, shall not be considered a change subject to this paragraph, but if that modification is thereafter to be regularly scheduled, the modified process shall be promptly filed as a scheduled process, accompanied by full information on the specified forms as provided in this paragraph.

§108.35, 2 (iii) Many packers employ an "operating" process in which retort operators are instructed to use retort temperatures and/or processing times slightly in excess of those specified in the scheduled process as a *safety factor* to compensate for minor fluctuations in temperature or time to assure that the minimum times and temperatures in the scheduled process are always met. ***This would not constitute a modification of the scheduled process.***

There are canning meat recipes that call for pasteurization only (212° F, 100° C or less), however, they are not included in the book as any error during processing will allow bacterial spores to survive and possibly grow. *Acidified* low-acid foods are usually pasteurized or processed at lower temperatures. However, the pasteurization of low-acid products should be left to commercial producers who are properly trained and equipped for this type of production. You might say that we are trying to scare you, (of course we are), that is our intention. People get sick and even die from eating canned vegetables or meats. Have you heard of anyone getting sick from eating orange marmalade? Even, if jam is covered with mold on top, most people just scoop it up (they should discard the jam) and keep on eating it. Here, we stress the point again: meats are not jams or jellies, they may be packed in the same size jars, but they conform to different processing rules.

Recipe Modifications

It has been mentioned a few times already that the recipes should be designed by *processing authority*. Occasionally, it may be difficult to follow exactly the prescribed process schedule, for example the recipe calls for a certain size of a container but we have a different one. Well, in extreme cases some substitution may be made, but always use common sense. It will be safe to *downsize* the size of the container (as long as the change will not adversely affect its cold point) keeping the same critical processing points such as the initial temperature, headspace, the thermal temperature and the processing time. For example, using 307 x 409 (No. 2, 21 oz can) instead of 404 x 414 (No. 3, 35 oz can). It is, however, dangerous to use a larger container or to increase the amount of starch or fat in the recipe.

Salt is added for the flavor and does not play safety role in the canning preservation method. Vegetables and meats can be canned successfully with or without salt.

Amount of Salt for Canning Vegetables and Meats	
Jar Size	Salt
Half Pint (8 oz)	1/4 tsp (1.5 g)
Pint (16 oz)	1/2 tsp (3 g)
Quart (32 oz)	1 tsp. (6 g)

Cured Meat/Poultry Products. Sodium nitrite (cure # 1, cure # 2) is the most effective agent that prevents *Cl. botulinum* spores from germinating. This is why it is always added to naturally smoked meats and sausages as the conditions in the smokehouse favor the growth of *Cl.botulinum*:

• Absence of air - we choke the air supply in order for the wood to start smoldering and smoking (having ample supply of air, the wood will burn cleanly without producing smoke).

• Moisture (meat contains 75% of water)

• Right temperature - most smoking is done between 86° and 140° F, 30-60° C.

Keep in mind that increasing the salt level enhances the inhibitory action of sodium nitrite. The conclusion is simple: using sodium nitrite with salt in a recipe will only increase the safety factor. This is the reason why many commercially produced canned meats are minimally heat processed.

Acidity. Acidity inhibits *Cl.botulinum* spores from germinating. Below pH 4.6 they will not germinate and such acidified low-acid foods can be processed at much lower temperatures. The conclusion: adding any amount of lemon juice, vinegar or citric acid will increase the safety factor. If you decide to do that, remember that you are not qualified to change the recipe, so maintain the original pressure, temperature and processing time.

Summary of Critical Issues

You will not jeopardize the safety of the recipe if you increase:

- Initial packing temperature.

- Cooking temperature.

- Cooking time. The quality and the texture of the product may be affected but the safety will only be increased.

- Cutting meat into smaller pieces and adding more liquid will result in faster heat penetration and increased safety.

- Adding starch or flour will decrease the safety as they absorb water and solidify during cooking what slows down heat transfer. After opening the can you can thicken the food with starch or flour before serving.

- Increasing the amount of fat or oil decreases the safety factor as fats act as heat insulators. They increase thermal resistance of bacterial spores by encapsulating them a thin layer of oil film.

- If you mix vegetables and meat you must process the mixture until both the meat and vegetables are safe to eat.

Recommended Containers for Home Use

Recommended Glass Jars for Home Use		
Common Name	Capacity	
	Fluid Ounces	Milliliters
1/2 Pint	8	236
Pint	16	473
Quart	32	946

Recommended Cans for Home Use			
Common Name	Size	Capacity	
1/2 pint (1 lb), tapered	307 x 200.25	8 oz	228 ml
Pint, tall salmon, tapered	301 x 408	16 oz	500 ml
No. 1	211 x 400	11 oz., (1⅓ cup)	310 ml
No. 2	307 x 409	20½ oz., (2½ cup)	600 ml
No. 2½, 30 oz	401 x 411	28 oz., (3½ cup)	850 ml
No. 3, quart	404 x 414	33 oz., (4 cup)	1000 ml

Containers larger than quart glass jars or No. 3 cans are not recommended for home use because of difficulty of thorough heat processing. As a result some parts of the food might not be sufficiently heated and may not kill bacterial spores.

However, larger cans can be used for packing dry foods (peas, beans, coffee beans) that don't have to be sterilized. Commercial producers can afford to use containers of any size or shape as the recipes they use are drawn by professionally trained individuals. In addition, those plants work under the direct supervision of Food Safety and Inspection Service inspectors and every recipe is thoroughly checked for safety. A home canner should follow USDA guidelines for canning food in round cans or glass jars.

There is very little information on processing times for metal cans as the USDA has stopped providing this information. University of Alaska in Fairbanks has a very comprehensive program in canning meats and fish, both in glass and metal containers. Pint and 1/2 pint cans are used as containers of choice.

Glass Jar	Volume	Metal Can	Volume
8 oz (1/2 Pint)	8.50 oz (251 ml)	307 x 200.25	7.75 oz (229 ml)
Pint	17.18 oz (508 ml)	301 x 408	16 oz (473 ml)
		307 x 409 (No. 2)	20.5 oz (606 ml)
Quart	33.95 oz (1003 ml)	401 x 411 (No. 2.5)	29.75 (879 ml)

In many recipes the same process schedule calls for either:

Pint jar or No. 2 can
Quart jar or No. 2.5 can

Being Safe

You may receive a home made canned product from a friend and the recipe and processing times may be unknown to you. To be extra safe, remove meat from the container, place in pan, add water, if needed; be certain meat is covered with water and, boil for 15 minutes. This will kill any vegetative bacteria and will deactivate toxin if present.

Altitude Adjustments. *Processing times and pressures are given for altitudes of 0-1000 feet for dial type and weighted gauge pressure canners.* If you are canning at higher altitudes, follow the USDA altitude adjustments listed below.

Canning Pressure (in pounds) at Different Altitudes		
Altitude (feet)	Dial Gauge Pressure Canner	Weighted Gauge Pressure Canner
0-1000	11	10
1001-2000	11	15
2001-4000	12	15
4001-6000	13	15
6001-8000	14	15

Note: make sure you know the type of your pressure canner.

If your recipe calls for 15 lbs. pressure at *sea level* increase the pressure 1 lb. for each 2000 feet altitude. Thus at an altitude of 4000 feet, process food at 17 lbs. of pressure instead of 15 lbs. pressure. If your canner will not allow you to increase the pressure over 15 lbs., increase processing time 20% for each 1,000 ft. rise in altitude.

Processing times for canning meats, poultry, fish and seafood in *glass jars* are adapted from the "Complete Guide to Home Canning," Agriculture Information Bulletin No. 539, USDA, revised 2009. *This guide includes processing times for glass jars only.* Those times and instructions come from USDA publication *AWI-110, "Home Canning of Meat"* that was printed in 1945, at the time of this publication (2013), they are still unchanged. The original publication also provided processing times for meat, poultry and fish for No. 2, No. 2½, and No. 3 tin cans. Those times agree with the data released in *AWI-110, USDA, 1945, "Home Canning of Meat"* and with the data released in 1946 in *Technical Bulletin No. 930 - "Home Canning Processes for Low-Acid Foods."*

Processing times for canning meats, poultry, fish and seafood in *cans* (No. 2, No. 2.5 and No. 3 cans) are quoted from:

AWI-110, USDA, 1945, "Home Canning of Meat."
Montana Extension Bulletin, 242, 1947, "Home Canning of Meat, Fish, Poultry."
USDA Home and Garden Bulletin No. 106, 1975 "Home Canning of Meat, Fish, Poultry."

Recipe Index

No	Recipe	Con-tainer	Page
	Meats		
1	Bacon in Aspic-Smoked	Can	217
2	Bacon with Peas	Jar	219
3	Bacon with Peas	Can	220
4	Beef in its Own Juice	Can, Jar	221
5	Beef Stew	Can, Jar	158
6	Beef With Buckwheat Groats	Can	223
7	Bigos (Hunter's Stew)	Can	224
8	Bigos (Hunter's Stew) with Tomatoes	Can	226
9	Chile con Carne	Jar	159
10	Corned Beef	Can, Jar	160
11	Festive Mincemeat Pie	Jar	161
12	Goulash	Can	228
13	Goulash Supreme	Can	230
14	Ground Meat	Can, Jar	162
15	Hamburger	Can, Jar	163
16	Heart and Tongue	Can, Jar	164
17	Lamb with Beans	Jar	231
18	Lamb with Rice	Jar	232
19	Luncheon Meat	Can	233
20	Lard-Homemade	Can	240
21	Lard-Plain	Can, Jar	241
22	Lard-Smoked-with Spices	Can, Jar	242
23	Lard with Onions	Can, Jar	243
24	Meat Stock (Broth, Soup Stock)	Can, Jar	165
25	Meat-Vegetable Stew	Can, Jar	166
26	Pate-Popular	Jar	244
27	Pate-Supreme	Can, Jar	245
28	Pork and Veal	Can	246
29	Pork-Ground	Can	247
30	Pork Hocks in Aspic	Can, Jar	248
31	Pork in its Own Juice	Can, Jar	250

32	Pork with Beans in Tomato Sauce	Can	252
33	Sausage	Can, Jar	167
34	Spaghetti Sauce with Meat	Jar	168
35	Strips, Cubes or Chunks of Meat	Can, Jar	169
36	Tripe Stew	Jar	253
37	Tripe Stew in Broth	Can, Jar	254
Poultry			
38	Chicken-Duck-Goose-Turkey & Game	Can, Jar	170
39	Chicken Soup Stock	Can, Jar	172
40	Giblets	Can, Jar	171
Game Meat			
41	Squirrel	Can, Jar	192
42	Rabbit	Can, Jar	191
Fish and Seafood			
43	Clams	Jar	176
44	Fish Balls	Can, Jar	177
45	Fish-Fried	Can, Jar	178
46	Fish Plain	Can, Jar	179
47	Fish in Pint Jars	Jar	180
48	Fish in Quart Jars	Jar	181
49	Fish-Smoked	Jar	183
50	Fish-Smoked	Can	184
51	Fish Fillets-Smoked	Can, Jar	185
52	King and Dungeness Crab Meat	Jar	186
53	Oysters	Jar	187
54	Shrimp-Gulf	Can, Jar	188
55	Tuna	Jar	189
Vegetables			
56	Asparagus-Spears or Pieces	Jar	194
57	Beans or Peas, Dried-Shelled	Jar	195
58	Beans-Baked	Jar	196
59	Beans, Dry, with Tomato or Molasses Sauce	Jar	197
60	Beans, Fresh Lima-Shelled	Jar	198
61	Beans, Snap and Italian, Pieces	Jar	199
62	Beets, Whole, Cubed or Sliced	Jar	200

63	Carrots, Sliced or Diced	Jar	201
64	Corn, Cream Style	Jar	202
65	Corn, Whole Kernel	Jar	203
66	Mixed Vegetables	Jar	204
67	Mushrooms, Whole or Sliced	Jar	205
68	Okra	Jar	206
69	Peas, Green or English, Shelled	Jar	207
70	Peppers	Jar	208
71	Potatoes-Sweet, Pieces or Whole	Jar	209
72	Potatoes-White, Cubed or Whole	Jar	210
73	Pumpkins and Winter Squash, Cubed	Jar	211
74	Soups	Jar	212
75	Spinach and Other Greens	Jar	213
76	Succotash	Jar	214

BEEF STEW

CFR 9, § 319.304 Meat stews.

Meat stews such as "Beef Stew" or "Lamb Stew" shall contain not less than 25 percent of meat of the species named on the label, computed on the weight of the fresh meat.

Beef, 300 g
Potatoes, 200 g
Peas, 50 g
Carrots, 120 g
Onions, 30 g
Salt, 10 g
Pepper, 20 g
Flour or starch, 20 g
Water or broth, 250 g

1. Peel potatoes, onions and carrots, then dice them. Keep freshly peeled potatoes under water to prevent darkening.
2. Trim beef, remove gristle and fat, cut meat into 1 inch (25 mm) cubes. Brown in a small amount of fat. Boil potatoes for 10 minutes. Mix flour with water, salt, and pepper to make gravy. Add hot meat and potatoes, vegetables and top with hot gravy leaving 1" headspace in glass jars and 1/4" headspace in cans. Remove air bubbles.
3. Adjust lids in glass jars and seal the cans. Process at once at 240° F, 116° C.

Beef Stew				
Style Pack	Container	Process Time	Canner Pressure at "0" ft	
			dial - gauge	weighted - gauge
Hot	Jar - Pint	75 min	11 lb	10 lb
	Jar - Quart	90 min		
	Can - No. 2	65 min		
For processing at above 1,000 ft, see page 154.				

CHILE CON CARNE

3 cups dried pinto or red kidney beans
5½ cups water
5 tsp salt (separated)
3 lbs ground beef
1½ cups chopped onions
1 cup chopped peppers of your choice (optional)
1 tsp black pepper
3 to 6 tbsp chili powder
2 quarts crushed or whole tomatoes
Yield: 9 pints

Procedure: Wash beans thoroughly and place them in a 2 qt. saucepan. Add cold water to a level of 2 to 3 inches above the beans and soak 12 to 18 hours. Drain and discard water. Combine beans with 5½ cups of fresh water and 2 teaspoons salt. Bring to a boil. Reduce heat and simmer 30 minutes. Drain and discard water. Brown ground beef, chopped onions, and peppers (if desired), in a skillet. Drain off fat and add 3 teaspoons of salt, pepper, chili powder, tomatoes and drained cooked beans. Simmer 5 minutes. **Caution:** *Do not thicken.* Fill hot jars, leaving 1 inch headspace. Remove air bubbles and adjust headspace if needed. Wipe rims of jars with a dampened clean paper towel. Adjust lids and process.

Chile Con Carne in *Glass Jars*				
Style of Pack	Jar Size	Process Time	Canner Pressure at "0" ft	
			dial-gauge	weighted-gauge
Hot	Pints	75 min	11 lb	10 lb
For processing at above 1,000 ft, see page 154.				*USDA 539, 2009.*

CORNED BEEF

Hot Pack

1. Wash the corned beef, cut into pieces suited for packing.

2. Cover meat with cold water and bring to boil. If broth tastes very salty, strain and cover meat with fresh water, and parboil again.

3. Pack hot meat. Leave about 1 inch above meat in glass jars for headspace. Leave 1-1/2 inch in cans.

4. Cover meat with hot broth or hot water. Leave 1 inch for headspace in jars, fill cans to top.

5. Work out bubbles with knife.

6. Seal and process at once.

Corned Beef				
Style Pack	Container	Process Time	Canner Pressure at "0" ft	
			dial - gauge	weighted - gauge
Hot	Jar - Pint	75 min	11 lb	10 lb
	Jar - Quart	90 min		
	Can - No. 2	65 min		
	Can - No. 2½ and No. 3	90 min		
For processing at above 1,000 ft, see page 154.				M 242, 1947.

FESTIVE MINCEMEAT PIE FILLING

2 cups finely chopped suet
4 lbs ground beef or (4 lbs ground venison and 1 lb sausage)
5 qts chopped apples
2 lbs dark seedless raisins
1 lb white raisins
2 qts apple cider
2 Tbsp ground cinnamon
2 tsp ground nutmeg
5 cups sugar
2 Tbsp salt
Yield: About 7 quarts

Procedure: cook suet and meat in water to avoid browning. Peel, core, and quarter apples. Put meat, suet, and apples through food grinder using a medium blade. Combine all ingredients in a large saucepan, and simmer 1 hour or until slightly thickened. Stir often. Fill jars with mixture without delay, leaving 1-inch headspace. Adjust lids and process.

Festive Mincemeat Pie Filling in *Glass Jars*				
Style of Pack	Jar Size	Process Time	Canner Pressure at "0" ft	
			dial-gauge	weighted-gauge
Hot	Quarts	90 min	11 lb	10 lb
For processing at above 1,000 ft, see page 154.			*USDA 539, 2009.*	

GROUND OR CHOPPED MEAT

Bear-Beef-Lamb-Pork-Sausage-Veal-Venison

Procedure: Choose fresh, chilled meat. For venison, add one part high-quality pork fat to three or four parts venison before grinding. Season freshly made sausage with salt and cayenne pepper (sage may cause a bitter off-flavor).

Hot pack: shape chopped meat into patties or balls or cut cased sausage into 3 to 4 inch links. Cook until lightly browned. Ground meat may be sauteed without shaping. Remove excess fat. Fill hot jars with pieces. Add boiling meat broth, tomato juice, or water, leaving 1 inch headspace. Remove air bubbles, adjust headspace if needed. Add 1 teaspoon of salt per quart, 1/2 teaspoon per pint, if desired. Wipe rims of jars with a dampened clean paper towel. Adjust lids and process.

Ground or Chopped Meat in *Glass Jars*				
Style of Pack	Jar Size	Process Time	Canner Pressure at "0" ft	
			dial-gauge	weighted-gauge
Hot	Pints	75 min	11 lb	10 lb
	Quarts	90 min		
For processing at above 1,000 ft, see page 154. *USDA 539, 2009.*				

Ground or Chopped Meat in Cans				
Style of Pack	Can Size	Process Time	Canner Pressure at "0" ft	
			dial-gauge	weighted-gauge
Hot	2	65 min	11 lb	10 lb
	2.5 and 3	90 min		
Raw *	2	100 min	11 lb	10 lb
	2.5 and 3	135 min		
For processing at above 1,000 ft, see page 154. *USDA, AWI-110,1945.*				

* Raw pack is suitable only for cans. It is difficult to remove ground meat out of glass jars when packed this way.

Raw Pack - *without forming cakes, pack raw ground meat solidly into cans level with top. Place cans in a pot with water about 2 inches below can rim. Exhaust (cook) at slow boil until meat registers 170° F, 77° C. Press meat down into cans leaving 1/2 inch headspace. Seal and process at once.*

HAMBURGER

Grind scraps of lean meat, or meat from less tender cuts. Season with 1 tsp. salt to 1 pound ground meat. Mix well.

Hot Pack

1. Form meat into small fairly thin, flat cakes, uniform in thickness (not humped in middle). Make them of a size to pack into the containers without breaking.
2. Place meat cakes in heavy baking pan, oiled slightly if necessary to prevent sticking. Precook in moderate oven until medium done. Cakes are heated enough when they show almost no red color at center when cut.
3. Pack cakes hot. Leave one-inch space above meat in glass jars; 1/2 inch in tin cans.
4. Cover meat with hot liquid (pan drippings with fat skimmed off, meat broth, or water). Again leave 1 inch space above liquid in glass jars. Fill tin cans to top.
5. Work out air bubbles. Add more liquid if necessary; leave 1 inch above liquid in glass jars, fill cans to top.
6. Seal and process at once.

Raw Pack * - Suitable only for *cans*.

1. Pack ground raw meat solidly into tin cans, level with the top.
2. Place open can in large vessel of water and heat as directed above for raw pack of cut pieces of meat.
3. Press meat down about 1/2 inch below rim of can. Seal can and process at once.

Hamburger				
Style of Pack	Container Size	Process Time	Canner Pressure at "0" ft	
			dial-gauge	weighted-gauge
Hot	Jar - Pint	75 min	11 lb	10 lb
	Jar - Quart	90 min		
	Can - No. 2	65 min		
	Can No. 2.5 and No. 3	90 min		
Raw *	Can - No. 2	100 min	11 lb	10 lb
	Can - No. 2.5 and 3	135 min		
For processing at above 1,000 ft, see page 154.				*M 242, 1947.*

HEART AND TONGUE

The heart and tongue are generally used as fresh meat. *If you do wish to can them follow directions for* beef, veal, pork, lamb as *hot packed.*

Procedure:

- Heart - remove thick connective tissue before cutting into pieces.

- Tongue - drop tongue into boiling water and simmer about 45 minutes or until skin can be removed, before cutting into pieces.

Hot Pack

1. Place hearts in boiling water. Simmer 15-20 minutes.

2. Reheat the prepared tongues in the broth or hot water.

3. Add 1 teaspoon of salt per quart to the jar, 1/2 teaspoon to pints, if desired.

4. Pack heart or tongue hot. Leave one inch headspace in glass jars, 1/2 inch in cans.

5. Cover meat with the hot pre-cooking liquid. Leave one inch space above liquid in glass jars, fill cans to top. Work out bubbles with a knife.

6. Seal and process at once at 10 pounds pressure (240° F, 116° C).

Heart and Tongue				
Style Pack	Container	Process Time	Canner Pressure at "0" ft	
			dial - gauge	weighted - gauge
Hot	Jar - Pint	75 min	11 lb	10 lb
	Can - No. 2	65 min		
For processing at above 1,000 ft, see page 154.				*M 242, 1947.*

MEAT STOCK (BROTH, SOUP STOCK)

Beef: Saw or crack fresh trimmed beef bones to enhance extraction of flavor. Rinse bones and place in a large stockpot or kettle, cover bones with water, add pot cover, and simmer 3 to 4 hours. Remove bones, cool broth, and pick off meat. Skim off fat, add meat trimmings removed from bones to broth, and reheat to boiling. Fill hot jars, leaving 1 inch headspace. Wipe rims of jars with a dampened clean paper towel. Adjust lids and process.

Chicken or turkey: Place large carcass bones (with most of meat removed) in a large stock pot, add enough water to cover bones, cover pot, and simmer 30 to 45 minutes or until remaining attached meat can be easily stripped from bones. Remove bones and pieces, cool broth, strip meat, discard excess fat, and return meat trimmings to broth. Reheat to boiling and fill jars, leaving 1 inch headspace. Wipe rims of jars with a dampened clean paper towel. Adjust lids and process.

Meat Stock in *Glass Jars*				
Style of Pack	Jar Size	Process Time	Canner Pressure at "0" ft	
			dial-gauge	weighted-gauge
Hot	Pints	20 min	11 lb	10 lb
	Quarts	25 min		
For processing at above 1,000 ft, see page 154.				*USDA 539, 2009.*

Meat Stock in Cans				
Style of Pack	Can Size	Process Time	Canner Pressure at "0" ft	
			dial-gauge	weighted-gauge
Hot	2	20 min	11 lb	10 lb
	2.5 and 3	25 min		
For processing at above 1,000 ft, see page 154.			*USDA, AWI-110, 1945.*	

MEAT - VEGETABLE STEW

Raw pack

Beef, lamb, or veal, cut in 1 -1/2-inch cubes - 2 quarts
Potatoes, pared or scraped, cut in 1/2 inch cubes - 2 quarts
Carrots, pared or scraped, cut in 1/2-inch cubes - 2 quarts
Celery, 1/4-inch pieces - 3 cups
Onions, small whole, peeled - 7 cups

Combine ingredients. Yield is 7 quarts or 16 pints.

Glass jars. Fill jars to top with raw meat-vegetable mixture. Add salt if desired: 1/2 teaspoon per pint or 1 teaspoon per quart. Adjust lids and process at once at 10 pounds pressure (240° F).

Metal cans. Fill cans to top with raw meat-vegetable mixture. Do not add liquid. Add salt if desired: 1/2 teaspoon to No. 2 cans or 1 teaspoon to No. 2.5 cans. To exhaust air, cook stew at slow boil to 170° F., or until medium done (about 50 minutes). Seal cans and process at once at 10 pounds pressure (240° F).

Meat - Vegetable Stew				
Style Pack	Container	Process Time	Canner Pressure at "0" ft	
			dial - gauge	weighted - gauge
Raw	Jar - Pint	60 min	11 lb	10 lb
	Jar - Quart	75 min		
	Can - No. 2	* 60 min		
	Can - No. 2.5	* 70 min		
For processing at above 1,000 ft, see page 154. *USDA 106, 1975.*				

* Original times were the following: No. 2 can - 40 min, No. 2.5 - 45 minutes. We have increased them to 60 minutes and 70 minutes.

SAUSAGE

Any good sausage recipe may be used. It is best to go easy on the spices because of flavor changes which may develop during canning and storing. Mold sausage into smooth, round flat cakes. Precook, pack and process the same as hamburger, *hot pack*.

The following recipe for sausage is suggested:

Use about one-third fat pork and two-thirds lean pork.
Mix seasonings and sprinkle over the meat before grinding.
For a fine, even sausage it should be run through the grinder a second, time.

For four pounds ground pork use:

5 level teaspoons salt
2 level teaspoons ground marjoram
1 level teaspoon sugar
1 level tablespoon pepper
1 level teaspoon ground cloves or 1/2 tsp. nutmeg, if desired.

The above amount may be made to test the recipe, if desired.

For 100 pounds ground pork use:

2-1/2 cups salt
1 cup ground marjoram
1/2 cup sugar (omit if it is expected that sausage will not
be used reasonably soon).
1-1/2 cups pepper

Sausage				
Style of Pack	Container Size	Process Time	Canner Pressure at "0" ft	
			dial-gauge	weighted-gauge
Hot	Jar - Pint	75 min	11 lb	10 lb
	Jar - Quart	90 min		
	Can - No. 2	65 min		
	Can No. 2.5 and No. 3	90 min		
For processing at above 1,000 ft, see page 154.				*M 242, 1947.*

SPAGHETTI SAUCE WITH MEAT

30 lbs tomatoes
2-1/2 lbs ground beef or sausage
5 cloves garlic, minced
1 cup chopped onions
1 cup chopped celery or green peppers
1 lb fresh mushrooms, sliced (optional)
4-1/2 tsp salt
2 tbsp oregano
4 tbsp minced parsley
2 tsp black pepper
1/4 cup brown sugar
Yield: About 9 pints

Procedure: do not increase the proportion of onions, peppers, or mushrooms.

1. Wash tomatoes and dip in boiling water for 30 to 60 seconds or until skins split. Dip in cold water and slip off skins. Remove cores and quarter tomatoes. Boil 20 minutes, uncovered, in large saucepan. Put through food mill or sieve.

2. Saute beef or sausage until brown. Add garlic, onion, celery or green pepper, and mushrooms, if desired. Cook until vegetables are tender.

3. Combine with tomato pulp in large sauce pan. Add spices, salt, and sugar. Bring to a boil. Simmer, uncovered, until thick enough for serving. At this time initial volume will have been reduced by nearly one-half. Stir frequently to avoid burning. Fill hot jars, leaving 1-inch headspace. Remove air bubbles and adjust headspace if needed. Wipe rims of jars with a dampened clean paper towel. Adjust lids and process.

Spaghetti Sauce With Meat				
Style of Pack	Jar Size	Process Time	Canner Pressure at "0" ft	
			dial-gauge	weighted-gauge
Hot	Pints	60 min	11 lb	10 lb
	Quarts	70 min		
For processing at above 1,000 ft, see page 154.			*USDA 539, 2009.*	

STRIPS, CUBES, OR CHUNKS OF MEAT

Bear-Beef-Lamb-Pork-Veal-Venison

Procedure: Choose quality chilled meat. Remove excess fat. Soak strong-flavored wild meats for 1 hour in brine water containing 1 tablespoon of salt per quart of water. Rinse. Remove large bones.

Hot pack - Precook meat until rare by roasting, stewing, or browning in a small amount of fat. Add 1 teaspoon of salt per quart to the jar, 1/2 teaspoon to pints, if desired. Fill hot jars with pieces and add boiling broth, meat drippings, water, or tomato juice (especially with wild game), leaving 1 inch headspace. Remove air bubbles and wipe jar rims.

Raw pack - Add 1 teaspoon of salt per quart to the jar, 1/2 teaspoon to pints, if desired. Pack raw meat in hot jars, leaving 1 inch headspace. Do not add liquid. Wipe rims of jars. Adjust lids and process.

Strips, Cubes, or Chunks of Meat in *Glass Jars*				
Style of Pack	Jar Size	Process Time	Canner Pressure at "0" ft	
			dial-gauge	weighted-gauge
Hot and Raw	Pints	75 min	11 lb	10 lb
	Quarts	90 min		
For processing at above 1,000 ft, see page 154. *USDA 539, 2009.*				

Strips, Cubes, or Chunks of Meat in *Cans*				
Style of Pack	Can Size	Process Time	Canner Pressure at "0" ft	
			dial-gauge	weighted-gauge
Hot and Raw	2	65 min	11 lb	10 lb
	2.5 and 3	90 min		
For processing at above 1,000 ft, see page 154. *USDA, AWI-110, 1945.*				

CANNED POULTRY

Chicken-Duck-Goose-Turkey & Game Birds

Procedure: Choose freshly killed and dressed, healthy animals. Large chickens are more flavorful than fryers. Dressed chicken should be chilled for 6 to 12 hours before canning. Remove excess fat. Cut the meat into suitable sizes for canning. Can with or without bones. Strong flavored game birds like water fowl may be soaked for 1 hour in a brine made from 1 tablespoon salt and 1 quart of water. If you soak game birds, don't add salt when packing jars.

Hot pack - Boil, steam, or bake meat until about two-thirds done. Add 1 tsp salt per quart, 1/2 tsp per pint, if desired. Fill hot jars with pieces and hot broth, leaving 1¼ inch headspace. Remove air bubbles.

Raw pack - Add 1 tsp salt per quart, 1/2 tsp per pint if desired. Fill hot jars loosely with raw meat pieces, leaving 1¼ inch headspace. Do not add liquid. Wipe rims of jars with a dampened clean paper towel. Adjust lids and process.

Poultry and Game Birds in *Glass Jars*				
Style of Pack	Jar Size	Process Time	Canner Pressure at "0" ft	
			dial-gauge	weighted-gauge
Hot and Raw, *Without* Bones	Pints	75 min	11 lb	10 lb
	Quarts	90 min		
Hot and Raw, *With* Bones	Pints	65 min	11 lb	10 lb
	Quarts	75 min		
For processing at above 1,000 ft, see page 154. *USDA 539, 2009.*				

Poultry and Game Birds in *Cans*				
Style of Pack	Can Size	Process Time	Canner Pressure at "0" ft	
			dial-gauge	weighted-gauge
Hot and Raw - *Without* Bones	2	65 min	11 lb	10 lb
	2.5 and 3	90 min		
Hot and Raw - *With* Bones	2	55 min	11 lb	10 lb
	2.5 and 3	75 min		
For processing at above 1,000 ft, see page 154. *USDA, AWI-110, 1945.*				

GIBLETS

Because of flavor, it is best to can livers alone. Gizzards and hearts, may be canned together. Since they are ordinarily canned and used in small quantities, directions are given only for pint glass jars and no. 2 cans.

Hot pack

1. Put giblets in cooking pan. Cover with broth made from bony pieces, or hot water. Cover pan and precook giblets until medium done. Stir occasionally.

2. If salt is desired, put level measure into clean, empty containers: 1/2 teaspoon in pint jar or No. 2 can.

3. Pack giblets hot. Leave about 1 inch above meat in glass jars for head space, 1/2 inch in cans.

4. Cover giblets with hot broth or hot water. Leave 1 inch for head space in jars, fill cans to top.

5. Work out air bubbles with knife. Add more liquid, if needed, to cover meat. Be sure to leave 1 inch headspace in jars and have cans filled to top.

6. Adjust lids on glass jars, seal cans.

7. Process at once in the steam pressure canner at 10 pounds pressure (240° F, 116° C).

Giblets				
Style Pack	Container	Process Time	Canner Pressure at "0" ft	
			dial - gauge	weighted - gauge
Hot	Jar - Pint	75 min	11 lb	10 lb
	Can - No. 2	65 min		
For processing at above 1,000 ft, see page 154. *USDA, AWI-110, 1945*				

SOUP STOCK

1. Make fairly concentrated stock by covering bony pieces of chicken or other meat with lightly salted water and simmering until meat is tender. Don't cook too long, or soup will loose flavor.

2. Skim off fat, remove all pieces of bone, but don't strain out meat and sediment.

3. Pour hot stock into containers. Leave 1 inch at top of glass for headspace; fill tin cans to top.

4. Adjust lids on glass jars, seal tin cans.

5. Process at once in pressure canner at 10 pounds pressure (240° F, 116° C).

Soup Stock				
Style Pack	Container	Process Time	Canner Pressure at "0" ft	
			dial - gauge	weighted - gauge
Hot	Jar - Pint	20 min	11 lb	10 lb
	Jar - Quart	25 min		
	Can - No. 2	20 min		
	Can - No. 2½ and No. 3	25 min		
For processing at above 1,000 ft, see page 154. *USDA, AWI-110, 1945.*				

CANNING FISH AND SEAFOOD

Caution: No container larger than a pint jar or No. 2 can should be used in the home canning of fish because difficulties in sterilization make the use of larger sized containers unsafe.

Read USDA recommendations for processing fish in quart glass jars that follow.

When canning halibut or other lean fish, up to 4 tablespoon of olive oil or vegetable oil may be added to each pint jar. The oil will add moisture to the product but will also increase the calories (1 Tbsp of oil = 135 cal).
Salt, seasoning salt, or other spices may be added to the packed fish.

Sauces

Sauces may be added to containers before sealing them or served with cooked canned fish.

Tomato Sauce. This sauce is often added to sardines, herrings, mackerel fillets and smoked fish. Tomato "fish sauce" is a type of ketchup made from tomato pulp, vinegar, onion, salt and spices. Bay leaves and cloves are often added. Sugar tends to caramelize at high temperatures so it should be added in small quantities as it may darken the sauce.

Tomato Sauce for Canned Fish

1 qt ripe whole tomatoes	1 medium onion
1 Tbsp chopped parsley	1 Tbsp vinegar
4 cloves	1 Tbsp Worcestersire sauce
1 bay leaf	1/4 tsp cayenne pepper
1/2 tsp salt	

Smash tomatoes and simmer with other ingredients until reduced to about half of the original volume. Strain through a fine strainer.

Mustard Sauce. This sauce is often added to sardines and other canned fish. Ingredients: mustard seed, vinegar, cayenne pepper, turmeric, salt, water.
Add 2 Tbsp vinegar to 1 cup of water. Add 5 Tbsp. cracked mustard seeds, 1/4 tsp cayenne, 1/2 tsp turmeric. Cook in skillet for 5 minutes. Strain.
Caution: Do not use starch, flour or any artificial thickening agent.

Canning Fish in 1/2 Pint and Pint Jars

Half-pint, pint or quart jars are suitable for canning fish. Fresh or frozen fish can be used, thaw frozen fish in the refrigerator. Prepare fish using general cooking procedures: bleed and gut fish immediately after catching. Make sure you remove gills. Remove the head, tail, fins and scales. You can leave bones in salmon and herring as they will soften during processing and storage. Remove the bones from halibut, when in doubt remove the bones.

Cut fish into fillets or chunks which are best suited for the jar or can that you are using. The skin may be left on or off. Fillets may be rolled for packing. Salmon, trout, mullet, herring are oily fish so no oil is needed, however lean fish like halibut or cod will benefit from additional fat or oil.

Follow the same procedures which are outlined for processing meat in glass jars.

Process 1/2 pint and pint jars for 100 minutes at:

10 psi - weighted gauge
11 psi - dial gauge

Canning fish in quart jars differs slightly.

Canning Fish in Quart Jars

When using quart sized jars, more time is required to heat the product thoroughly. Heat the canner on high for 20 minutes until steam comes through vent pipe in a steady stream. Allow the steam to escape for 10 minutes to vent the canner. The total time it takes to *heat and vent* the canner filled with quart jars should never be less than 30 minutes. The total time may be more than 30 minutes, especially if you have tightly packed jars, cold fish, or a very large canner. Close the vent, bring the canner up to the recommended pressure and process containers:

Fish in quarts jars - 160 minutes at:

10 psi - weighted gauge
11 psi - dial gauge

Canning Fish in Cans

Fresh or frozen fish can be used, thaw frozen fish in the refrigerator. Prepare fish using general cooking procedures: bleed and gut fish immediately after catching. Make sure you remove gills. Remove the head, tail, fins and scales. You can leave bones in salmon and herring as they will soften during processing and storage. Remove the bones from halibut, when in doubt remove the bones.

Cut fish into fillets or chunks which are best suited for the jar or can that you are using. The skin may be left on or off. Fillets may be rolled for packing. Salmon, trout, mullet, herring are oily fish so no oil is needed, however lean fish like halibut or cod will benefit from additional fat or oil.

Cans and lids: 1 pound (size: 301 × 408) or ½ pound (size: 307 × 200.25), also called Alaska salmon cans. According to University of Alaska in Fairbanks 25 pounds of fish as caught will fill 12, 1-pound cans or 24, ½-pound cans.

Follow procedure for canning meats in can, however note that fish are exposed to *different (longer) processing times.*

Process one pound cans (301 x 408) for 115 minutes at:

10 psi - weighted gauge
11 psi - dial gauge

Process half pound cans (307 x 200.25) for 95 minutes at:

10 psi - weighted gauge
11 psi - dial gauge

CLAMS

Whole or minced.

Hot Pack - keep clams live on ice until ready to can. Scrub shells thoroughly and rinse, steam 5 minutes, and open. Remove clam meat. Collect and save clam juice. Wash clam meat in water containing 1 teaspoon of salt per quart. Rinse and cover clam meat with boiling water containing 2 tablespoons of lemon juice or ½ teaspoon of citric acid per gallon. Boil 2 minutes and drain. To make minced clams, grind clams with a meat grinder or food processor. Fill *hot* jars *loosely* with pieces and add *hot* clam juice and *boiling water* if needed, leaving *1 inch headspace*. Remove air bubbles and adjust headspace if needed. Wipe rims of jars with a dampened clean paper towel. Adjust lids and process.

Clams in *Glass Jars*				
Style of Pack	Jar Size	Process Time	Canner Pressure at "0" ft	
			dial-gauge	weighted-gauge
Hot	Half-pints	60 min	11 lb	10 lb
	Pints	70 min		
For processing at above 1,000 ft, see page 154. *USDA 539, 2009.*				

FISH BALLS

Use fillets and all fish meat that can be scraped off the bones.

Grind meat through 3 mm/1/8" plate.

Ground fish, 1.5 kg/3.5 lb

Milk, 1 cup
Fish broth*, 1 cup
Flour, 3 Tbsp
Fish sauce, 5 Tbsp
Nutmeg, 1 tsp
Pepper, 2 tsp
Ginger, 1/3 tsp

* simmer fish bones, in little water for 60 minutes.

Mix all ingredients together, then add ground fish and remix again.

Form the mixture into balls about 2 inch thick.

Fry the balls in hot oil (375° F, 190° C) until light brown. This should not take longer than 1 minute. Pack into containers and fill with hot fish broth or light tomato sauce. Leave 1" headspace in glass jars, and 1/4" headspace in cans.

Seal and process at once.

Fish Balls				
Style Pack	Container	Process Time	Canner Pressure at "0" ft	
			dial - gauge	weighted - gauge
Hot	Jar - Pint	100 min	11 lb	10 lb
	Can - No. 2	90 min		
For processing at above 1,000 ft, see page 154.				

FISH - FRIED

1. Fish for canning should be strictly fresh. It is well to stick fish with a knife to drain out the blood as soon as they are caught. The viscera should also be removed as soon as possible.

2. Scale or wash carefully. Scales are easier to remove if fish are dipped into boiling water and then into cold water. If skins are tough, skin the fish.

3. In order to draw out the blood before canning, place fish in a brine made by using 1 tablespoon salt to each quart of water. Let fish soak 10 minutes to one hour, according to thickness. Small trout need only 10 minutes' soaking. This soaking is not absolutely essential but makes for a better looking product. It tends to make fish firmer.

4. Remove from brine, wash and drain on clean towels.

5. Leave whole or cut into convenient, uniform pieces.

6. Season fish before frying if they were not soaked in brine.

7. Fry in deep fat till nicely browned. Drain and place on brown paper to remove excess fat.

8. Pack into clean hot jars. Add about 2 tablespoon hot liquid. Add hot tomato sauce if desired and in that case use glass jars or R-enameled cans.

9. Seal and process at once.

Tomato Sauce for Canned Fish

1 qt. canned tomatoes
1 T chopped parsley
3 or 4 whole cloves
Few drops Tabasco sauce
1 medium onion
Piece of bay leaf
1 tablespoon Worcestershire sauce

Cook all ingredients until reduced to about half the original volume. Put through a fine strainer. This sauce may be put over fish in the jars before processing or served with fresh cooked or plain canned fish.

Fried Fish				
Style Pack	Container	Process Time	Canner Pressure at "0" ft	
			dial - gauge	weighted - gauge
Hot	Jar - Pint	100 min	11 lb	10 lb
	Can - No. 2	90 min		
For processing at above 1,000 ft, see page 154.				*M 242, 1943.*

FISH - PLAIN

An excess quantity of fish caught in spring and summer may be canned for use in other seasons. Fish for canning must be strictly fresh. Fish deteriorates rapidly in quality and flavor, so quick work is essential for a high grade product.

Procedure:

1. Fish for canning should be strictly fresh. It is well to stick fish with a knife to drain out the blood as soon as they are caught. The viscera should also be removed as soon as possible.
2. Scale or wash carefully. Scales are easier to remove if fish are dipped into boiling water and then into cold water. If skins are tough, skin the fish.
3. In order to draw out the blood before canning, place fish in a brine made by using 1 tablespoon salt to each quart of water. Let fish soak 10 minutes to one hour, according to thickness. Small trout need only 10 minutes' soaking. This soaking is not absolutely essential but makes for a better looking product. It tends to make fish firmer.
4. Pack fish into clean hot jars or tin cans, packing in "up and down" or "circular" fashion to make a good looking jar. Large fish will need to be cut into convenient, uniform pieces.
5. Add 1 teaspoon salt per pint jar if fish have not been previously soaked in brine. Do not add water. Hot tomato sauce may be poured over the fish in the jars or enameled cans.
6. Exhaust tin cans till steaming hot. Seal and process at once.

Note: Fish canned in this manner will resemble canned salmon in texture.

Plain Fish				
Style Pack	Container	Process Time	Canner Pressure at "0" ft	
			dial - gauge	weighted - gauge
Hot	Jar - Pint	100 min	11 lb	10 lb
	Can - No. 2	90 min		
For processing at above 1,000 ft, see page 154.				*M 242, 1943.*

FISH IN PINT JARS

Blue, mackerel, salmon, steelhead, trout, and other fatty fish except tuna.

Caution: Bleed and eviscerate fish immediately after catching, never more than 2 hours after they are caught. Keep cleaned fish on ice until ready to can.

Note: Glass-like crystals of struvite, or magnesium ammonium phosphate, sometime form in canned salmon. There is no way for the home canner to prevent these crystals from forming, but they usually dissolve when heated and are safe to eat.

Procedure: If the fish is frozen, thaw it in the refrigerator before canning. Rinse the fish in cold water. You can add vinegar to the water (2 tablespoons per quart) to help remove slime. Remove head, tail, fins, and scales; it is not necessary to remove the skin. You can leave the bones in most fish because the bones become very soft and are a good source of calcium. For halibut, remove the head, tail, fins, skin, and the bones. Wash and remove all blood. Refrigerate all fish until you are ready to pack in jars. Split fish lengthwise, if desired. Cut cleaned fish into 3-1/2-inch lengths. If the skin has been left on the fish, pack the fish skin out, for a nicer appearance or skin in, for easier jar cleaning. Fill hot pint jars, leaving 1-inch headspace. Add 1 teaspoon of salt per pint, if desired. Do not add liquids. Carefully clean the jar rims with a clean, damp paper towel; wipe with a dry paper towel to remove any fish oil. Adjust lids and process. Fish in half-pint or 12-ounce jars would be processed for the same amount of time as pint jars.

Fish in *Glass Jars*				
Style of Pack	Jar Size	Process Time	Canner Pressure at "0" ft	
			dial-gauge	weighted-gauge
Raw	Pints	100 min	11 lb	10 lb
For processing at above 1,000 ft, see page 154.				*USDA 539, 2009.*

FISH IN QUART JARS

Blue, mackerel, salmon, steelhead, trout, and other fatty fish except tuna.

Note: Glass-like crystals of struvite, or magnesium ammonium phosphate, sometime form in canned salmon. There is no way for the home canner to prevent these crystals from forming, but they usually dissolve when heated and are safe to eat.

Caution: Bleed and eviscerate fish immediately after catching, never more than 2 hours after they are caught. Keep cleaned fish on ice until ready to can.

Procedure: If the fish is frozen, thaw it in the refrigerator before canning. Rinse the fish in cold water. You can add vinegar to the water (2 tablespoons per quart) to help remove slime. Remove head, tail, fins, and scales; it is not necessary to remove the skin. You can leave the bones in most fish because the bones become very soft and are a good source of calcium. For halibut, remove the head, tail, fins, skin, and the bones. Wash and remove all blood. Refrigerate all fish until you are ready to pack in jars. Cut the fish into jar-length filets or chunks of any size. The one-quart straight-sided mason type jar is recommended. If the skin has been left on the fish, pack the fish skin out, for a nicer appearance or skin in, for easier jar cleaning. Pack solidly into hot quart jars, leaving *1 inch headspace.* If desired, run a plastic knife around the inside of the jar to align the product; this allows firm packing of fish. For most fish, no liquid, salt, or spices need to be added, although seasonings or salt may be added for flavor (1 to 2 teaspoons salt per quart, or amount desired). *For halibut, add up to 4 tablespoons of vegetable or olive oil per quart jar if you wish.* The canned product will seem moister. However, the oil will increase the caloric value of the fish. Carefully clean the jar rims with a clean, damp paper towel; wipe with a dry paper towel to remove any fish oil. Adjust lids and process.

Processing Change for Quart Jars: The directions for operating the pressure canner during processing of quart jars are different from those for processing pint jars, so please read the following carefully. It is critical to product safety that the processing directions are followed exactly. When you are ready to process your jars of fish, *add 3 quarts of water to the pressure canner.* Put the rack in the bottom of canner and place closed jars on the rack. Fasten the canner cover securely, but do not close the lid vent. Heat the canner on high for 20 minutes. If steam comes through the open vent in a steady stream at the end of 20 minutes, allow it to escape for an additional 10 minutes. If steam does not come through the open vent in a steady stream at the end of 20 minutes, keep heating the canner until it does. Then allow the steam to escape for an additional 10 minutes to vent the canner. This step removes air from inside the canner so the temperature is the same throughout the canner.

The total time it takes to heat and vent the canner should never be less than 30 minutes. The total time may be more than 30 minutes if you have tightly packed jars, cold fish, or larger sized canners. For safety's sake, *you must have a complete, uninterrupted 160 minutes* (2 hours and 40 minutes) at a minimum pressure required for your altitude. Write down the time at the beginning of the process and the time when the process will be finished.

Fish in *Glass Jars*				
Style of Pack	Jar Size	Process Time	Canner Pressure at "0" ft	
			dial-gauge	weighted-gauge
Raw	Quarts	160 min	11 lb	10 lb
For processing at above 1,000 ft, see page 154.				*USDA 539, 2009.*

FISH - SMOKED

Salmon, rockfish and flatfish (sole, cod, flounder) and other fish

Caution: Safe processing times for other smoked seafoods have not been determined. Those products should be frozen. Smoking of fish should be done by tested methods. Lightly smoked fish is recommended for canning because the smoked flavor will become stronger and the flesh drier after processing. However, because it has not yet been cooked, do not taste lightly smoked fish before canning. Follow these recommended canning instructions carefully. *Use a 16 to 22 quart pressure canner* for this procedure; do not use smaller pressure saucepans. Safe processing times have not been determined. *Do not use jars larger than one pint.* Half-pints could be safely processed for the same length of time as pints, but the quality of the product may be less acceptable.

Procedure: If smoked fish has been frozen, thaw in the refrigerator until no ice crystals remain before canning. If not done prior to smoking, cut fish into pieces that will fit vertically into pint canning jars, leaving *1 inch headspace.* Pack smoked fish vertically into hot jars, leaving *1 inch headspace* between the pieces and the top rim of the jar. The fish may be packed either loosely or tightly.

Do not add liquid to the jars. Clean jar rims with a clean, damp paper towel. Adjust lids and process.

Processing Change for Smoked Fish: The directions for filling the pressure canner for processing smoked fish are different than those for other pressure canning, so please read the following carefully. It is critical to product safety that the processing directions are followed exactly. When you are ready to process your jars of smoked fish, measure 4 quarts (16 cups) of *cool* tap water and pour into the pressure canner. Note: The water level probably will reach the screw bands of pint jars. *Do not decrease the amount of water or heat the water before processing begins.* Place prepared, closed jars on the rack in the bottom of the canner, and proceed as with usual pressure canning instructions.

Smoked Fish in *Glass Jars*				
Style of Pack	Jar Size	Process Time	Canner Pressure at "0" ft	
			dial-gauge	weighted-gauge
see instructions above	Pints	110 min	11 lb	10 lb
For processing at above 1,000 ft, see page 154.				*USDA 539, 2009.*

All fish can be smoked but the oily fish such as salmon, trout, herring, mackerel, eel, blue fish, and mullet taste the best.

FISH - SMOKED IN CANS

Gut and clean the fish. Soak in 20% brine for 60 minutes and then smoke at 100 F, 38 C for 90 minutes.

Cut the smoked fish to fit the cans.

Pack the fish solidly leaving 1/4 inch headspace.

Exhaust. Place a suitable pot or pan on the stove with some hot water in it. Have more hot water nearby.

Place filled with fish *open* cans in the hot water in a single layer. The water should come up halfway up the outside of the cans. Do not cover cans with the lids. Covering the pot will result in moisture dripping into the cans.

Turn heat on high and bring water to boil. Adjust the heat to keep a steady boil.

Heat the pot until the fish reach a temperature of 170° F, 77° C. Using a jar lifter remove cans from the pot and seal one can at a time.

Place sealed cans on the rack in the canner. The canner should be already filled with about 3 inches of hot water. Check your canner specifications.

Process at once.

Smoked Fish Fillets				
Style of Pack	Container	Process Time	Canner Pressure at "0" ft	
			dial-gauge	weighted-gauge
Hot	1/2 lb can, 307 x 200.25	110 min	11 lb	10 lb
	1 lb can, 301 x 408	125 min		
For processing at above 1,000 ft, see page 154.				

FISH FILLETS - SMOKED

Mullet, mackerel or other fish.

Gut and clean the fish. Soak in 20% brine for 60 minutes and then fillet.

Lay fillet on screens and smoke at 100° F, 38° C for 90 minutes.

Glass Jars

Pack warm smoked fillets solidly into the jars. Fillets can be rolled before packing. Leave one inch of headspace between the fish and the top of the jar. Place 2-3 inches of warm water in the canner, and the rack and place the jars. Process at once.

Metal Cans

Exhaust. Place a suitable pot or pan on the stove with some hot water in it. Have more hot water nearby.

Place warm smoked fillets solidly into cans. Leave one inch headspace. Place the *open* cans in the boiling water in a single layer. The water should come up halfway up the outside of the cans. Do not cover cans with the lids.

Heat the uncovered pot until the fish reach a temperature of 170° F, 77° C. Covering the pot will result in moisture dripping into the cans.

Remove cans from the pot and seal. Process at once.

Smoked Fish Fillets				
Style of Pack	Container	Process Time	Canner Pressure at "0" ft	
			dial-gauge	weighted-gauge
Hot	Jar - 1/2-pints	100 min	11 lb	10 lb
	Cans - 307 x 200.25	95 min		
For processing at above 1,000 ft, see page 154.				*USDA 539, 2009.*

Note: no additional fat or oil is needed for canning oily fish like mullet, mackerel, salmon or herring. When canning lean fish, up to 2 tablespoons of olive/vegetable oil may be added to 1/2 pint jar or 1/2 lb can.

KING AND DUNGENESS CRAB MEAT

It is recommended that blue crab meat be frozen instead of canned for best quality. Crab meat canned according to the following procedure may have a distinctly acidic flavor and freezing is the preferred method of preservation at this time.

Procedure: Keep live crabs on ice until ready to can. Wash crabs thoroughly, using several changes of cold water. Simmer crabs 20 minutes in water containing cup of lemon juice and 2 tablespoons of salt (or up to 1 cup of salt, if desired) per gallon. Cool in cold water, drain, remove back shell, then remove meat from body and claws. Soak meat 2 minutes in cold water containing 2 cups of lemon juice or 4 cups of white vinegar, and 2 tablespoons of salt (or up to 1 cup of salt, if desired) per gallon. Drain and squeeze crab meat to remove excess moisture. Fill *hot* half-pint jars with 6 ounces of crab meat and pint jars with 12 ounces, leaving *1 inch headspace.* Add ½ teaspoon of citric acid or 2 tablespoons of lemon juice to each half-pint jar, or 1 teaspoon of citric acid or 4 tablespoons of lemon juice per pint jar. Cover with fresh *boiling water,* leaving *1 inch headspace.* Remove air bubbles and adjust headspace if needed. Wipe rims of jars with a dampened clean paper towel. Adjust lids and process.

King and Dungeness Crab Meat in *Glass Jars*				
Style of Pack	Jar Size	Process Time	Canner Pressure at "0" ft	
			dial-gauge	weighted-gauge
Hot	Half-pints	70 min	11 lb	10 lb
	Pints	80 min		
For processing at above 1,000 ft, see page 154. *USDA 539, 2009.*				

OYSTERS

Procedure: Keep live oysters on ice until ready to can. Wash shells. Heat 5 to 7 minutes in preheated oven at 400°F (204° C). Cool briefly in ice water. Drain, open shell, and remove meat. Wash meat in water containing ½ cup salt per gallon. Drain. Add ½ teaspoon salt to each pint, if desired. Fill *hot* half-pint or pint jars with drained oysters and cover with fresh *boiling water*, leaving *1 inch headspace*. Remove air bubbles and adjust headspace if needed. Wipe rims of jars with a dampened clean paper towel. Adjust lids and process.

Oysters in *Glass Jars*				
Style of Pack	Jar Size	Process Time	Canner Pressure at "0" ft	
			dial-gauge	weighted-gauge
see instructions above	Half-pints or pints	75 min	11 lb	10 lb
For processing at above 1,000 ft, see page 154.			*USDA 539, 2009.*	

SHRIMP - Gulf

Peel the shrimp from their shells.

Wash the shrimp and place into 10% (40° salinometer) brine for 10 minutes. Stir the shrimp after 5 minutes. To make brine dissolve 1 lb of salt in 1 gallon of water.

Remove shrimp and drop them into a fresh boiling 10% brine for 6 minutes. Drain and place on the screen until the surface moisture evaporates away, then pack into containers without a delay.

Glass Jars

Pack shrimp into into glass jars to within 1-1/4 from the top. About 5 ounces of shrimp are put into 8 oz (1/2 pint) jar.

Fill glass jars with a hot 1% brine. To make brine dissolve 0.70 oz (20 g = 3 teaspoons) salt in 1 gallon of water. Leave 1 inch headspace. Seal and process at once.

Cans

Pack shrimp into cans to within 1/2" from the top.

Fill glass jars with a hot 1% brine. Leave 1/4 inch headsets. To make brine dissolve 0.35 oz (10 g) salt in 1 gallon of water. Seal and process at once.

Shrimp-Gulf				
Style of Pack	Container	Process Time	Canner Pressure at "0" ft	
			dial-gauge	weighted-gauge
Hot	Jar - 1/2-pints	25 min	11 lb	10 lb
	Jar - 1 pint	35 min		
	Can - 307 x 200.25	20 min		
	Can - No. 2	30 min		
For processing at above 1,000 ft, see page 154.				

Shrimp are often packed into the No. 1 picnic, 211 x 400, can and processed for 20 minutes at 240° F, 116° C.

Glass packed shrimp should be kept out of sunlight to prevent discoloration.

TUNA

Tuna may be canned *either precooked or raw*. Precooking removes most of the strong-flavored oils. The strong flavor of dark tuna flesh affects the delicate flavor of white flesh. Many people prefer not to can dark flesh. It may be used as pet food.

Note: glass-like crystals of struvite, or magnesium ammonium phosphate, sometime form in canned tuna. There is no way for the home canner to prevent these crystals from forming, but they usually dissolve when heated and are safe to eat.

Procedure: Keep tuna on ice until ready to can. Remove viscera and wash fish well in cold water. Allow blood to drain from stomach cavity. Place fish belly down on a rack or metal tray in the bottom of a large baking pan. Cut tuna in half crosswise, if necessary. Precook fish by baking at 250°F (121° C) for 2½ to 4 hours (depending on size) or at 350°F (177° C) for 1 hour. The fish may also be cooked in a steamer for 2 to 4 hours. If a thermometer is used, cook to a 165° to 175°F (74 - 80° C) internal temperature. Refrigerate cooked fish overnight to firm the meat. Peel off the skin with a knife, removing blood vessels and any discolored flesh. Cut meat away from bones; cut out and discard all bones, fin bases, and dark flesh. Quarter. Cut quarters crosswise into lengths suitable for half-pint or pint jars. Fill into hot jars, pressing down gently to make a solid pack. *Tuna may be packed in water or oil,* whichever is preferred. Add water or oil to jars, leaving *1 inch headspace*. Remove air bubbles and adjust headspace if needed. Add ½ teaspoon of salt per half-pint or 1 teaspoon of salt per pint, if desired. Carefully clean the jar rims with a clean, damp paper towel; wipe with a dry paper towel to remove any fish oil. Adjust lids and process.

Tuna in *Glass Jars*				
Style of Pack	Jar Size	Process Time	Canner Pressure at "0" ft	
			dial-gauge	weighted-gauge
see instructions above	Half-pints or pints	100 min	11 lb	10 lb
For processing at above 1,000 ft, see page 154.				*USDA 539, 2009.*

GAME MEAT

Wild animals such as boar, moose, raccoon, bear are often infected with a parasite that causes a disease called Trichinae. The pigs that roam free and are not on controlled diet can also be infected. The parasite is, however, easily destroyed by cooking pork to 140° F, 60° C, although it is recommended to cook wild game to 165° F, 74° C. It is inadvisable to eat such meats raw. Deer which does not eat meat is an exception and venison is free of the Trichinalis parasite. The parasite will not survive canning temperature so there is no danger.

Game meat will benefit from brief soaking in a solution of salt and water (brine).

Canned wild meat also benefits by covering it with a soup stock broth or tomato - based broth. Boullion cube will add extra flavor if the stock is weak.

Spices such as rosemary and juniper berries are often used with game meat.

RABBIT

Cut rabbit into suitable sizes for canning. Soak meat for 1 hour in water containing 1 tablespoon of salt per quart, and rinse. Prepare the meaty pieces, with or without bones, and pack and process as for poultry, omitting the salt.

Hot pack - Boil, steam, or bake meat until about two-thirds done. Add 1 tsp salt per quart, 1/2 tsp per pint, if desired. Fill hot jars with pieces and hot broth, leaving 1¼ inch headspace. Remove air bubbles.

Raw pack - Add 1 tsp salt per quart, 1/2 tsp per pint if desired. Fill hot jars loosely with raw meat pieces, leaving 1¼ inch headspace. Do not add liquid. Wipe rims of jars with a dampened clean paper towel. Adjust lids and process.

Rabbit in *Glass Jars*				
Style of Pack	Jar Size	Process Time	Canner Pressure at "0" ft	
			dial-gauge	weighted-gauge
Hot and Raw, *Without* Bones	Pints	75 min	11 lb	10 lb
	Quarts	90 min		
Hot and Raw, *With* Bones	Pints	65 min	11 lb	10 lb
	Quarts	75 min		
For processing at above 1,000 ft, see page 154. *USDA 539, 2009.*				

Rabbit in *Cans*				
Style of Pack	Can Size	Process Time	Canner Pressure at "0" ft	
			dial-gauge	weighted-gauge
Hot and Raw - *Without* Bones	2	65 min	11 lb	10 lb
	2.5 and 3	90 min		
Hot and Raw - *With* Bones	2	55 min	11 lb	10 lb
	2.5 and 3	75 min		
For processing at above 1,000 ft, see page 154. *USDA, AWI-110, 1945*				

SQUIRREL

Soak meat for 1 hour in water containing 1 tablespoon of salt per quart, and then rinse. *Use processing times for poultry*, omitting the salt.

Hot pack - Boil, steam, or bake meat until about two-thirds done. Add 1 tsp salt per quart, 1/2 tsp per pint, if desired. Fill hot jars with pieces and hot broth, leaving 1¼ inch headspace. Remove air bubbles.

Raw pack - Add 1 tsp salt per quart, 1/2 tsp per pint if desired. Fill hot jars loosely with raw meat pieces, leaving 1¼ inch headspace. Do not add liquid. Wipe rims of jars with a dampened clean paper towel. Adjust lids and process.

Squirrel in *Glass Jars*				
Style of Pack	Jar Size	Process Time	Canner Pressure at "0" ft	
			dial-gauge	weighted-gauge
Hot and Raw, *Without* Bones	Pints	75 min	11 lb	10 lb
	Quarts	90 min		
Hot and Raw, *With* Bones	Pints	65 min	11 lb	10 lb
	Quarts	75 min		
For processing at above 1,000 ft, see page 154.				

Squirell in *Cans*				
Style of Pack	Can Size	Process Time	Canner Pressure at "0" ft	
			dial-gauge	weighted-gauge
Hot and Raw - *Without* Bones	2	65 min	11 lb	10 lb
	2.5 and 3	90 min		
Hot and Raw - *With* Bones	2	55 min	11 lb	10 lb
	2.5 and 3	75 min		
For processing at above 1,000 ft, see page 154.				

Vegetables

The following table is quoted from Putra University, Malaysia which has a very strong food science program.

Product	Initial Temperature		No. 2 can, 307 x 409	
	° F	° C	Minutes at 240° F/116° C	Minutes at 250° F/121° C
Green Beans	70	21	21	12
French Beans	70	21	40	20
Corn, cream	160	71	100	80
Corn, whole	100	38	55	30
Peas	70	21	36	16
Pumpkin	160	71	80	65

USDA vegetable canning recipes are provided for glass jars only. To replace a jar with a can, consider the following:

- Follow exactly USDA recipes, including canner pressure settings and processing times. Do not add more sugar, syrup, sauces, starches, fats or oils. Those ingredients will affect processing times, making it harder to kill bacterial spores.

- Use Hot Pack method only. Precooking vegetables opens up the cell structure and allows air to escape. During the heating process the air in the glass jar can escape through the sealant, but once the can is sealed the air has no place to go. The air inhibits vacuum creation and the transfer of heat.

- After packing vegetables, fill the can with hot boiling water leaving 1/4 inch headspace and process at once.

- Substitute glass jars with cans of smaller volume.

ASPARAGUS - SPEARS OR PIECES

Use tender, tight-tipped spears, 4 to 6 inches long.

Procedure: Wash asparagus and trim off tough scales. Break off tough stems and wash again. Cut into 1-inch pieces or can whole.

Hot pack - Cover asparagus with boiling water. Boil 2 or 3 minutes. Loosely fill hot jars with

hot asparagus, leaving 1-inch headspace.

Raw pack - Fill hot jars with raw asparagus, packing as tightly as possible without crushing, leaving 1-inch headspace.

Add 1 teaspoon of salt per quart to the jars, if desired. Add boiling water, leaving 1-inch headspace. Remove air bubbles and adjust headspace if needed. Wipe rims of jars with a dampened clean paper towel. Adjust lids and process.

Asparagus in *Glass Jars*				
Style of Pack	Jar Size	Process Time	Canner Pressure at "0" ft	
			dial-gauge	weighted-gauge
Hot and Raw	Pints	30 min	11 lb	10 lb
	Quarts	40 min		
For processing at above 1,000 ft, see page 154. *USDA 539, 2009.*				

BEANS OR PEAS - DRIED, SHELLED

All varieties

Select mature, dry seeds. Sort out and discard discolored seeds.

Procedure: Place dried beans or peas in a large pot and cover with water. Soak 12 to 18 hours in a cool place. Drain water. To quickly hydrate beans, you may cover sorted and washed beans with boiling water in a saucepan. Boil 2 minutes, remove from heat, soak 1 hour and drain. Cover beans soaked by either method with fresh water and boil 30 minutes.

Add 1/2 teaspoon of salt per pint or 1 teaspoon per quart to the jar, if desired. Fill hot jars with beans or peas and cooking water, leaving 1-inch headspace. Remove air bubbles, adjust lids and process.

Beans or Peas - Shelled, Dried in *Glass Jars*				
Style of Pack	Jar Size	Process Time	Canner Pressure at "0" ft	
			dial-gauge	weighted-gauge
Hot	Pints	75 min	11 lb	10 lb
	Quarts	90 min		
For processing at above 1,000 ft, see page 154.			*USDA 539, 2009.*	

BEANS, BAKED

Procedure: Soak and boil beans and prepare molasses sauce.

Molasses Sauce - Mix 4 cups water or cooking liquid from beans, 3 tablespoons dark molasses, 1 tablespoon vinegar, 2 teaspoons salt, and 3/4 teaspoon powered dry mustard. Heat to a boiling.

Place seven 3/4-inch pieces of pork, ham, or bacon in an earthenware crock, a large casserole, or a pan. Add beans and enough molasses sauce to cover beans. Cover and bake 4 to 5 hours at 350° F, 177° C. Add water as needed-about every hour. Fill hot jars, leaving 1-inch headspace. Remove air bubbles, adjust lids and process.

Beans - Baked in *Glass Jars*				
Style of Pack	Jar Size	Process Time	Canner Pressure at "0" ft	
			dial-gauge	weighted-gauge
Hot	Pints	65 min	11 lb	10 lb
	Quarts	75 min		
For processing at above 1,000 ft, see page 154. *USDA 539, 2009.*				

Note: the thermal process for beans in sauce depends in part upon the sauce formulation. Changing the sauce formula will affect the processing time.

BEANS, DRY, WITH TOMATO OR MOLASSES SAUCE

Select mature, dry seeds. Sort out and discard discolored seeds.

Procedure: Sort and wash dry beans. Add 3 cups of water for each cup of dried beans or peas. Boil 2 minutes, remove from heat and soak 1 hour and drain. Heat to boiling in fresh water, and save liquid for making sauce. Make your choice of the following sauces:

Tomato Sauce 1 - Mix 1 quart tomato juice, 3 tablespoons sugar, 2 teaspoons salt, 1 tablespoon chopped onion, and 1/4 teaspoon each of ground cloves, allspice, mace, and cayenne pepper. Heat to boiling.

Tomato Sauce 2 - Mix 1 cup tomato ketchup with 3 cups of cooking liquid from beans. Heat to boiling.

Molasses Sauce - Mix 4 cups water or cooking liquid from beans, 3 tablespoons dark molasses, 1 tablespoon vinegar, 2 teaspoons salt, and 3/4 teaspoon powered dry mustard. Heat to boiling.

Fill hot jars three-fourths full with hot beans. Add *one* 3/4-inch cube of pork, ham, or bacon to each jar, if desired. Fill jars with heated sauce, leaving 1-inch headspace. Remove air bubbles, adjust lids and process.

Beans, dry, with Tomato or Molasses sauce in *Glass Jars*				
Style of Pack	Jar Size	Process Time	Canner Pressure at "0" ft	
			dial-gauge	weighted-gauge
Hot	Pints	65 min	11 lb	10 lb
	Quarts	75 min		
For processing at above 1,000 ft, see page 154.				USDA 539, 2009.

Note: the thermal process for beans in sauce depends in part upon the sauce formulation. Changing the sauce formula will affect the processing time.

Do not add more meat or bacon.

BEANS, FRESH LIMA - SHELLED

Select well-filled pods with green seeds. Discard insect-damaged and diseased seeds.

Procedure: Shell beans and wash thoroughly.

Hot pack - Cover beans with boiling water and heat to boil. Fill hot jars loosely, leaving 1-inch headspace. Add 1/2 teaspoon of salt to pints, 1 teaspoon to quarts, if desired. Fill with boiling water to 1 inch from top. Remove air bubbles, adjust lids and process.

Raw pack - Fill hot jars loosely with raw beans. Do not press or shake down.
Small beans - leave 1-inch of headspace for pints and 1-1/2 inches for quarts.
Large beans - leave 1-inch of headspace for pints and 1-1/4 inches for quarts.
Add 1/2 teaspoon of salt to pints, 1 teaspoon to quarts, if desired. Add boiling water, leaving the same headspaces listed above. Remove air bubbles, adjust lids and process.

Beans - Lima in *Glass Jars*				
Style of Pack	Jar Size	Process Time	Canner Pressure at "0" ft	
			dial-gauge	weighted-gauge
Hot and Raw	Pints	40 min	11 lb	10 lb
	Quarts	50 min		
For processing at above 1,000 ft, see page 154.				*USDA 539, 2009.*

BEANS, SNAP AND ITALIAN - PIECES

Green and wax

Select filled but tender, crisp pods. Remove and discard diseased and rusty pods.

Procedure: Wash beans and trim ends. Leave whole or cut or snap into 1-inch pieces.

Hot pack - Cover with boiling water; boil 5 minutes. Fill hot jars, loosely leaving 1-inch headspace. Add 1/2 teaspoon of salt to pints, 1 teaspoon to quarts, if desired. Fill with boiling water to 1 inch from top. Remove air bubbles, adjust lids and process.

Raw pack - Fill hot jars tightly with raw beans, leaving 1-inch headspace.

Add 1/2 teaspoon of salt to pints, 1 teaspoon to quarts, if desired. Add boiling water, leaving 1-inch headspace. Remove air bubbles, adjust lids and process.

Beans, Snap and Italian, Pieces in *Glass Jars*				
Style of Pack	Jar Size	Process Time	Canner Pressure at "0" ft	
			dial-gauge	weighted-gauge
Hot and Raw	Pints	20 min	11 lb	10 lb
	Quarts	25 min		
For processing at above 1,000 ft, see page 154.			*USDA 539, 2009.*	

BEETS - WHOLE, CUBED, OR SLICED

Beets with a diameter of 1 to 2 inches are preferred for whole packs. Beets larger than 3 inches in diameter are often fibrous.

Procedure: Cut off beet tops, leaving an inch of stem and roots to reduce loss of color. Scrub well. Cover with boiling water. Boil until skins slip off easily; about 15 to 25 minutes depending on size. Cool, remove skins, and trim off stems and roots. Leave baby beets whole. Cut medium or large beets into 1/2-inch cubes or slices. Halve or quarter very large slices. Pack beets into hot jars leaving 1 inch headspace. Add 1/2 teaspoon of salt to pints, 1 teaspoon to quarts, if desired. Fill jars with hot water to 1 inch from top. Remove air bubbles, adjust lids and process.

Beets Whole, Cubed, or Sliced in *Glass Jars*				
Style of Pack	Jar Size	Process Time	Canner Pressure at "0" ft	
			dial-gauge	weighted-gauge
Hot	Pints	30 min	11 lb	10 lb
	Quarts	35 min		
For processing at above 1,000 ft, see page 154. *USDA 539, 2009.*				

CARROTS - SLICED OR DICED

Select small carrots, preferably 1 to 1-1/4 inches in diameter. Larger carrots are often too fibrous.

Procedure: Wash, peel, and rewash carrots. Slice or dice.

Hot pack - Cover carrots with boiling water; bring to boil and simmer for 5 minutes. Fill hot jars, leaving 1-inch of headspace. Add 1/2 teaspoon of salt to pints, 1 teaspoon to quarts, if desired. Fill jars with hot water to 1 inch from top.

Raw pack - Fill hot jars tightly with raw carrots, leaving 1-inch headspace.

Add 1/2 teaspoon of salt to pints, 1 teaspoon to quarts, if desired. Add hot cooking liquid or water, leaving 1-inch headspace. Remove air bubbles and adjust headspace if needed. Wipe rims of jars with a dampened clean paper towel. Adjust lids and process.

Carrots Sliced or Diced in *Glass Jars*				
Style of Pack	Jar Size	Process Time	Canner Pressure at "0" ft	
			dial-gauge	weighted-gauge
Hot and Raw	Pints	25 min	11 lb	10 lb
	Quarts	30 min		
For processing at above 1,000 ft, see page 154.				*USDA 539, 2009.*

CORN - CREAM STYLE

Select ears containing slightly immature kernels, or of ideal quality for eating fresh.

Procedure: Husk corn, remove silk, and wash ears. Blanch ears 4 minutes in boiling water. Cut corn from cob at the center of kernel. Scrape remaining corn from cobs with a table knife.

Hot pack - Add 1 cup of boiling water to 2 cups of corn.
Heat to a boil. Fill hot pint jar with hot corn, leaving 1-inch headspace. Add 1/2 teaspoon salt to each pint jar, if desired. Remove air bubbles, adjust lids and process.

Corn - Cream Style in *Glass Jars*				
Style of Pack	Jar Size	Process Time	Canner Pressure at "0" ft	
			dial-gauge	weighted-gauge
Hot	Pints	85 min	11 lb	10 lb
For processing at above 1,000 ft, see page 154. *USDA 539, 2009.*				

CORN - WHOLE KERNEL

Select ears containing slightly immature kernels or of ideal quality for eating fresh. Canning of some sweeter varieties or too immature kernels may cause browning. Can a small amount, check color and flavor before canning large quantities.

Procedure: Husk corn, remove silk, and wash. Blanch 3 minutes in boiling water. Cut corn from cob at about 3/4 the depth of kernel. *Caution:* Do not scrape cob.

Hot pack - To each clean quart of kernels in a saucepan, add 1 cup of hot water, heat to boiling and simmer 5 minutes. Fill hot jars with corn and cooking liquid, leaving 1-inch headspace. Add 1/2 teaspoon of salt to pints, 1 teaspoon to quarts, if desired. Fill jars with hot water to 1 inch from top. Remove air bubbles, adjust lids and process.

Raw pack - Fill hot jars with corn, leaving 1-inch headspace. Add 1/2 teaspoon of salt to pints, 1 teaspoon to quarts, if desired. Fill jars with hot water to 1 inch from top. Remove air bubbles, adjust lids and process.

Whole Corn in *Glass Jars*				
Style of Pack	Jar Size	Process Time	Canner Pressure at "0" ft	
			dial-gauge	weighted-gauge
Hot and Raw	Pints	55 min	11 lb	10 lb
	Quarts	85 min		
For processing at above 1,000 ft, see page 154.			*USDA 539, 2009.*	

MIXED VEGETABLES

6 cups sliced carrots
6 cups cut, whole kernel sweet corn
6 cups cut green beans
6 cups shelled lima beans
4 cups whole or crushed tomatoes
4 cups diced zucchini
Yield: 7 quarts

Optional mix - You may change the suggested proportions or substitute other favorite vegetables except leafy greens, dried beans, cream-style corn, squash and sweet potatoes.

Procedure: Except for zucchini, wash and prepare vegetables for canning. Wash, trim, and slice or cube zucchini (if used); combine all vegetables in a large pot or kettle, and add enough water to cover pieces. Boil 5 minutes.

Hot Pack - fill hot jars with hot pieces, leaving 1-inch headspace. Add 1/2 teaspoon of salt to pints, 1 teaspoon to quarts, if desired. Fill jars with hot water to 1 inch from top. Remove air bubbles, adjust lids and process.

Mixed Vegetables in *Glass Jars*				
Style of Pack	Jar Size	Process Time	Canner Pressure at "0" ft	
			dial-gauge	weighted-gauge
Hot	Pints	75 min	11 lb	10 lb
	Quarts	90 min		
For processing at above 1,000 ft, see page 154.				*USDA 539, 2009.*

MUSHROOMS - WHOLE OR SLICED

Select only brightly colored, small to medium-size domestic mushrooms with short stems, tight veils (unopened caps), and no discoloration.

Caution: Do not can wild mushrooms.

Procedure: Trim stems and discolored parts. Soak in cold water for 10 minutes to remove dirt. Wash in clean water. Leave small mushrooms whole; cut large ones.

Hot Pack - Cover with water in a saucepan and boil 5 minutes. Fill hot jars with hot mushrooms, leaving 1-inch headspace. Add 1/4 teaspoon of salt to half-pints, 1/2 teaspoon to pints, if desired. For better color, add 1/8 teaspoon of ascorbic acid powder, or a crushed 500-milligram tablet of vitamin C per pint. Fill jars with hot water to 1 inch from top. Remove air bubbles, adjust lids and process.

Mushrooms, Whole or Sliced in *Glass Jars*				
Style of Pack	Jar Size	Process Time	Canner Pressure at "0" ft	
			dial-gauge	weighted-gauge
Hot	1/2 Pints or Pints	45 min	11 lb	10 lb
For processing at above 1,000 ft, see page 154.				*USDA 539, 2009.*

OKRA

Select young, tender pods. Remove and discard diseased and rust-spotted pods.

Procedure: Wash pods and trim ends. Leave whole or cut into 1-inch pieces.

Hot Pack - Cover with hot water in a saucepan, boil 2 minutes and drain. Fill hot jars with hot okra and cooking liquid, leaving 1-inch headspace. Add 1/2 teaspoon of salt to pints, 1 teaspoon to quarts, if desired. Remove air bubbles, adjust lids and process.

Okra in *Glass Jars*				
Style of Pack	Jar Size	Process Time	Canner Pressure at "0" ft	
			dial-gauge	weighted-gauge
Hot	Pints	25 min	11 lb	10 lb
	Quarts	40 min		
For processing at above 1,000 ft, see page 154. *USDA 539, 2009.*				

PEAS, GREEN OR ENGLISH - SHELLED

It is recommended that sugar snap and Chinese edible pods be frozen for best quality. Select filled pods containing young, tender, sweet seeds. Discard diseased pods.

Procedure: Shell and wash peas.

Hot pack - Cover with boiling water. Bring to a boil in a saucepan, and boil for 2 minutes. Fill hot jars loosely with hot peas, and add cooking liquid, leaving 1-inch headspace. Add 1/2 teaspoon of salt to pints, 1 teaspoon to quarts, if desired. Remove air bubbles, adjust lids and process.

Raw pack - Fill hot jars with raw peas, add boiling water, leaving 1-inch headspace. Do not shake or press down peas. Add 1/2 teaspoon of salt to pints, 1 teaspoon to quarts, if desired. Remove air bubbles, adjust lids and process.

Peas, Green or English in *Glass Jars*				
Style of Pack	Jar Size	Process Time	Canner Pressure at "0" ft	
			dial-gauge	weighted-gauge
Hot and Raw	Pints or Quarts	40 min	11 lb	10 lb
For processing at above 1,000 ft, see page 154.			*USDA 539, 2009.*	

Note: For dry peas follow procedures and processing times for dry beans.

PEPPERS

Hot or sweet, including chiles, jalapeño, and pimiento.

Procedure: Select firm yellow, green, or red peppers. Do not use soft or diseased peppers.

Caution: If you choose hot peppers, wear plastic or rubber gloves and do not touch your face while handling or cutting hot peppers. If you do not wear gloves, wash hands thoroughly with soap and water before touching your face or eyes. If you do touch your eyes you may have vision problems for 15-20 minutes, however, the condition is not threatening and the problem will go away. Remember, hot peppers are the main ingredient of self-defence pepper spray.

Hot Pack - Small peppers may be left whole. Large peppers may be quartered. Remove cores and seeds. Slash two or four slits in each pepper, and either blanch in boiling water or blister skins using one of these two methods:
1. Oven or broiler method to blister skins – place peppers in a hot oven (400°F, 204° C) or broiler for 6-8 minutes until skins blister.
2. Range-top method to blister skins – Cover hot burner, either gas or electric, with heavy wire mesh. Place peppers on burner for several minutes until skins blister. After blistering skins, place peppers in a pan and cover with a damp cloth. (This will make peeling the peppers easier).
Cool several minutes; peel off skins. Flatten whole peppers. Add 1/2 teaspoon of salt to each pint jar, if desired. Fill hot jars loosely with peppers and add fresh boiling water, leaving 1-inch headspace. Remove air bubbles, adjust lids and process.

Peppers in *Glass Jars*				
Style of Pack	Jar Size	Process Time	Canner Pressure at "0" ft	
			dial-gauge	weighted-gauge
Hot	1/2 Pints or Pints	35 min	11 lb	10 lb
For processing at above 1,000 ft, see page 154. *USDA 539, 2009.*				

POTATOES, SWEET - PIECES OR WHOLE

Choose small to medium-sized potatoes. They should be mature and not too fibrous. Can within 1 to 2 months after harvest.

Hot Pack - Wash potatoes and boil or steam until partially soft (15 to 20 minutes). Remove skins. Cut medium potatoes, if needed, so that pieces are uniform in size.
Caution: Do not mash or puree pieces.

Fill hot jars, leaving 1-inch headspace. Add 1/2 teaspoon of salt to pints, 1 teaspoon to quarts, if desired. Fill jars with hot water boiling water or syrup to 1 inch from top.
Medium Syrup (30%) - 5 cups sugar, 1-3/4 quart water.
Remove air bubbles, adjust lids and process.

Potatoes, Sweet in *Glass Jars*				
Style of Pack	Jar Size	Process Time	Canner Pressure at "0" ft	
			dial-gauge	weighted-gauge
Hot	Pints	65 min	11 lb	10 lb
	Quarts	90 min		
For processing at above 1,000 ft, see page 154.				*USDA 539, 2009.*

POTATOES, WHITE - CUBED OR WHOLE

Select small to medium-size mature potatoes of ideal quality for cooking. Tubers stored below 45°F may discolor when canned Choose potatoes 1 to 2 inches in diameter if they are to be packed whole.

Procedure: Wash and peel potatoes. If desired, cut into 1/2-inch cubes. To prevent darkening place potatoes in ascorbic acid solution. Drain.

Ascorbic acid solution: add 3000 mg (1 teaspoon) of ascorbic acid powder, or 6 crushed 500-milligram tablet of vitamin C to 1 gallon of water.

Hot Pack - Cook 2 minutes in boiling water and drain again. For whole potatoes, boil 10 minutes and drain. Fill hot jars with hot potatoes leaving 1-inch headspace. Add 1/2 teaspoon of salt to pints, 1 teaspoon to quarts, if desired. Fill jars with boiling water to 1 inch from top. Remove air bubbles Adjust lids and process.

White Potatoes in *Glass Jars*				
Style of Pack	Jar Size	Process Time	Canner Pressure at "0" ft	
			dial-gauge	weighted-gauge
Hot	Pints	35 min	11 lb	10 lb
	Quarts	40 min		
For processing at above 1,000 ft, see page 154. *USDA 539, 2009.*				

PUMPKINS AND WINTER SQUASH - CUBED

Pumpkins and squash should have a hard rind and stringless, mature pulp of ideal quality for cooking fresh. Small size pumpkins (sugar or pie varieties) make better products.

Procedure: Wash, remove seeds, cut into 1-inch-wide slices, and peel. Cut flesh into 1-inch cubes. Boil 2 minutes in water.

Caution: Do not mash or puree.

Hot Pack - Fill hot jars with cubes and boiling hot cooking liquid, leaving 1-inch headspace. Remove air bubbles, adjust lids and process.

For making pies, drain jars and strain or sieve the cubes at preparation time.

Pumpkin and Winter Squash in *Glass Jars*				
Style of Pack	Jar Size	Process Time	Canner Pressure at "0" ft	
			dial-gauge	weighted-gauge
Hot	Pints	55 min	11 lb	10 lb
	Quarts	90 min		
For processing at above 1,000 ft, see page 154. *USDA 539, 2009.*				

SQUASH, WINTER - CUBED

Prepare and process according to instructions for "Pumpkin" above.

SOUPS

Vegetable, dried bean, pea, meat, poultry, or seafoods.

Caution: Do not add noodles or other pasta, rice, flour, cream, milk or other thickening agents to home canned soups. If dried beans or peas are used, they must be fully rehydrated first.

Procedure: Select, wash, and prepare vegetables, meat, and seafoods as described for the specific foods. Cover meat with water and cook until tender. Cool meat and remove bones. Cook vegetables. For each cup of dried beans or peas, add 3 cups of water, boil 2 minutes, remove from heat, soak 1 hour, and heat to boil.
Drain all foods and combine with meat broth, tomatoes, or water to cover. Boil 5 minutes.

Caution: Do not thicken. Salt to taste, if desired.

Hot Pack - Fill hot jars only halfway with mixture of solids. Add and cover with remaining liquid, leaving 1-inch headspace. Remove air bubbles, adjust lids and process.

Soups in *Glass Jars*				
Style of Pack	Jar Size	Process Time	Canner Pressure at "0" ft	
			dial-gauge	weighted-gauge
Hot	Pints	60* min	11 lb	10 lb
	Quarts	75* min		
* Caution: process 100 minutes if soup contains seafoods.				
For processing at above 1,000 ft, see page 154. *USDA 539, 2009.*				

SPINACH AND OTHER GREENS

Can only freshly harvested greens. Discard any wilted, discolored, diseased, or insect-damaged leaves. Leaves should be tender and attractive in color.

Procedure: Wash only small amounts of greens at one time. Drain water and continue rinsing until water is clear and free of grit. Cut out tough stems and midribs. Place 1 pound of greens at a time in cheesecloth bag or blancher basket and steam 3 to 5 minutes or until well wilted. Add 1/4 teaspoon of salt to pints and 1/2 teaspoon of salt to quarts, if desired.

Hot Pack - Fill hot jars loosely with greens and add fresh boiling water, leaving 1-inch headspace. Remove air bubbles, adjust lids and process.

Spinach and Other Greens in *Glass Jars*				
Style of Pack	Jar Size	Process Time	Canner Pressure at "0" ft	
			dial-gauge	weighted-gauge
Hot	Pints	70 min	11 lb	10 lb
	Quarts	90 min		
* Caution: process 100 minutes if soup contains seafoods.				
For processing at above 1,000 ft, see page 154. *USDA 539, 2009.*				

SUCCOTASH

15 lbs unhusked sweet corn or 3 qts cut whole kernels
14 lbs mature green podded lima beans or 4 qts shelled limas
2 qts crushed or whole tomatoes (optional)
Yield: 7 quarts

Procedure: Wash and prepare fresh produce as described previously for specific vegetables.

Hot pack - Combine all prepared vegetables in a large kettle with enough water to cover the pieces. Boil succotash gently 5 minutes and fill hot jars with pieces and hot cooking liquid, leaving 1-inch headspace. Add 1/2 teaspoon of salt to pints and 1 teaspoon of salt to quarts, if desired.

Raw pack - Fill hot jars with equal parts of all prepared vegetables, leaving 1-inch headspace. Do not shake or press down pieces. Add 1/2 teaspoon of salt to pints and 1 teaspoon of salt to quarts, if desired. Fill with fresh boiling water to 1 inch from top.
Remove air bubbles, adjust lids and process.

Succotash in *Glass Jars*				
Style of Pack	Jar Size	Process Time	Canner Pressure at "0" ft	
			dial-gauge	weighted-gauge
Hot and Raw	Pints	60 min	11 lb	10 lb
	Quarts	85 min		
* Caution: process 100 minutes if soup contains seafoods.				
For processing at above 1,000 ft, see page 154. *USDA 539, 2009.*				

Polish Canning Recipes

The following are Polish commercial canning recipes which were produced by Polish plants for sale to consumers. Between 1945 to 1990 the communist system controlled the country, the industry and commerce. In simple terms everything belonged to the government. The canning plants were not an exception and similar to the sausage industry, whatever product they produced had to conform to the government regulations. That included recipes which were prepared by the meat technologists and the plants had to follow those instructions down to the letter. The recipes which follow are scaled down versions of the original instructions which were calculated for 100 kg (220 lb) of material, the amount that is too large to process at home.

You will notice that the meat is often listed as pork grade I, pork grade II or pork grade III. This does not mean that grade I is superior to grade III as all fresh meat is good. Meat plants don't select meats using definitions as butt, belly, picnic, or ham. In order to remove noble cuts of meat as ham, loin, or bacon, the meat cutter has to trim a significant amount of smaller cuts of meat, some with fat, others with a lot of connective tissue such as sinews, gristle, silver skin. He separates those cuts as:

Pork grade I - very lean meat.
Pork grade II - meat with some fat.
Pork grade III - meat with fat and connective tissue.
Pork grade IV - traces of blood, tendons, glands.

This system is not limited to pork only, but beef is similarly classified. Some cuts, for example pork butt and picnic contain pork of all four grades.

Pork grade III and skins are of great interest as they bind water and are rich in collagen, the material that upon heating will produce the natural gelatin. This also produces a pleasant mouthfeel. Meat containing connective tissue is ground finely or emulsified in order to magnify its binding properties.

You cannot produce a good quality sausage with the lean meat alone and it is impossible to produce liver or head cheese unless meats rich in collagen as well as the skins are used. So, if you see pork meat grade III, this means that the meat with a lot of connective tissue was chosen in order to produce the high quality product. The same explanation is applied to pork skins.

Exhausting Cans

It is important that as much air as possible is removed from the cans.

Place the canner rack on the bottom of canner. Exhaust only one layer of cans at a time. Place *open*, filled cans inside the water in the canner. Add enough water to come *half way* up the sides of the *open* cans filled with meat.

If you have more than will fit in one layer, put the second layer in a *different pan and heat it separately*. Place cans in the water where the liquid comes about halfway up the sides. Set the open cans, filled with meat, in a single layer in simmering water in open roasting pans on the stove. Do not cover the canner or the pan, otherwise condensing moisture will drop into the cans.

Adjust the temperature so the water comes to a *gentle boil*. Check the temperature of the meat in the cans with a meat thermometer. The internal temperature of the meat in the center of the can must reach 170° F, 77° C. The meat will loose most of its raw color. This process usually requires 30-50 minutes for beef and less for chicken.

Remove cans from the boiling water using a jar lifter. Carefully clean the edges of each can with a towel.

After exhausting syrups, brine, meat stock should be added to containers as hot as possible.

BACON IN ASPIC-SMOKED

Materials

Bacon, skinless or not, 1 kg (2.2 lb)
Use the freshest bacon possible (no previously frozen).

Ingredients for 1 kg (2.2 lb) of material

Pepper, 0.5 g (1/4 tsp)
Marjoram, 0.1 g (1/10 tsp)
Bay leaf, one leaf, finely crushed
Allspice, 0.5 g (1/4 tsp)
Gelatin*, 5 g (1 tsp)
For skinless bacon, double the amount of gelatin (10 g)

Curing Instructions

Dry mix:

Salt, 25 g (4 tsp)
Cure #1, 1.5 g (1/4 tsp)
Sugar, 10 g (2 tsp)

1. Cut bacon slabs onto sizes that will fit reasonably tight into your curing container.
2. Rub in 1/2 dry mix into all areas of bacon and place the slabs skin down for 48 hours. Cure at 40 F°, 4° C.
3. Restack the bacon slabs: top ones go on the bottom and the bottom ones on top. Apply the remaining 1/2 of curing solution when restacking the pieces. Cure for additional 48 hours at 40 F°, 4° C.
4. Make a hanging loop on one end and hang bacon for 2-3 hours in a cooler to drain and dry. When the surface feels dry, proceed to smoking.
5. Place bacon in a preheated to 122° F, 50° C, smokehouse and when bacon feels dry, apply warm smoke (86-104° F, 30-40° C) for about 2 hours until the skin becomes yellow.
6. Start preparing liquid stock broth before smoking bacon.

Option I - take 350 ml (12 fl oz) of water and bring to a boil. Immerse a spice bag with pepper, marjoram, bay leaf and allspice in hot water. This liquid will be added to meat in the can. Gelatin powder will be directly added to the can.

Option II - make a strong stock from pork legs or skins. Cover legs/ skins with a little water and simmer for 2-3 hours. Strain the liquid and place in refrigerator to set. Discard the fat from the top.

Place a bag spice with all spices inside in 100 ml (3.4 oz) of water, bring to a boil and simmer for 20 minutes. Mix this spice liquid with 250 ml (1 cup) of the pork legs/skins stock. This liquid bacon stock will be added hot to cans. Don't use gelatin when adding this stock.

7. Pour gelatin powder (if used) on the bottom of the can.
8. Cut bacon into strips that will fit the height of the can best (remember about headspace). They may be rolled (wrapped) around. Add 820 g (28 oz) of bacon, you should not have no more than 3 principal pieces. If more bacon is needed, cut a strip size that complement one of the main strips.
9. Pack the bacon into the cans leaving 1/2" headspace. Add broth to 1/2 inch from top.
10. Exhaust cans to 170° F, 77° C meat temperature.
11. Stir carefully removing the air from the bottom and sides, then add more hot liquid to 1/4" from top. Seal immediately and process at once at 240 F, 116 C:

No. 2 can (307 x 409) - 60 min.
No. 2.5 can (401 x 411 mm) - 65 min.

Notes

* Instead of using commercial gelatin, you may prepare your own stock from pig feet or pork skins (remove all fat).

BACON WITH PEAS in Jars

Materials

Smoked bacon, 300 g (0.66 lb)
Lard, 10 g (2 tsp)
Dry peas, 690 g (13.76 oz)

Ingredients for 1 kg (2.2 lb) of material

Salt, 16 g (2-1/2 tsp)
Pepper, 0.5 g (1/4 tsp)
Marjoram, 0.5 g (1/4 tsp)
1/2 Small onion, 20 g (3/4 oz)
Allspice, 0.5 g (1/4tsp)
Paprika, 0.5 g (1/4 tsp)
Bay leaf, 1/2
Leek, 10 g (1/3 oz)
Celery, 10 g (1/3 oz)
Carrot, 10 g (1/3 oz)
Parsley, 10 g, (1/3 oz)
Pork bones for making stock, 150 g (6 oz)

Instructions

1. Soak dry peas in water for 3 hours. Stir.
2. Dice the onion and fry in lard until golden.
3. Cut bacon into pieces about 2-1/2 oz (70 g) each.
4. Add 1 quart (1 liter) of water to a skillet and add bones. Bring to a boil and slow boil for 3 hours. Add all ingredients and boil additional hour (total boil 4 hours). Strain the stock and mix with fried onions.
5. Place in each jar: 1-2 pieces of bacon, peas and fill with hot stock leaving 1 inch headspace. Seal and process at once at 250° F, 121° C:

1/2 Pint jars - 50 min.
Pint jars - 60 min.

BACON WITH PEAS in Cans

Materials

Smoked bacon, 600 g (1.32 lb)
Lard, 10 g (2 tsp)
Dry peas, 390 g (13.76 oz)

Ingredients for 1 kg (2.2 lb) of material

Salt, 5 g (1 tsp)
Marjoram, 0.5 g (1/4 tsp)
Small onion, 40 g (1.4 oz)

Instructions

1. Soak dry peas in cold water for 12 hours. Stir often.
2. Add water to pot equipped with a steamer basked and bring to a boil. Place soaked peas in a steamer basket and cook for 3 minutes, shaking the pot a few times to make sure that all peas are scalded equally. Drain peas on screens.
3. Dice onion and fry in lard until golden.
4. Cook smoked bacon for 3 minutes in boiling water. Cut bacon into pieces about 2-1/2 oz (70 g) each.
5. Mix peas with onion, salt and marjoram. Add peas and mix again.
6. Pack each can with one piece of smoked bacon, then fill with peas.
7. Add water, leaving 1/2 headspace.
8. Exhaust cans to 170° F, 77° C meat temperature. Fill with hot water to 1/4 inch from top.
9. Seal and process at once at 250° F, 121° C:

307 x 200.25 - 45 min.

BEEF IN ITS OWN JUICE

Materials

Beef, grade I, II or III, 840 g
Beef grade IV, 40 g
Beef head meat*, 20 g
Pork skins, 80 g
Tallow (beef fat), melted, 20 g
* substitution with other lean beef is allowed.

Ingredients for 1 kg (2.2 lb) of material

Salt, 18 g (3 tsp)
Pepper, 1 g (1/2 tsp)
Paprika, 0.5 g (1/4 tsp)
Marjoram, 0.2 g
Ginger, 0.1 g
Onion, 10 g (1/4 small onion)

Instructions

1. Clean heads from any remaining glands and bloody areas.
2. Skins.
a. Wash raw skins in lukewarm water.
b. Cure skins for 2 days in 18° Be (70° salinometer, 18% salt) at refrigerator temperature. Use enough solution to be able to mix the skins without difficulty.
70° SAL solution: 1.88 lb salt to 1 gallon of water.
c. Simmer the skins for 25 minutes at 194-203° F, 90-95° C, but don't make them too soft.
d. Cool the skins on screens as warm skins stick together during grinding.

Note: the skins may be substituted with pork or beef hocks, meat with a lot of connective tissue, for example pork III grade from picnic (front leg) or pig ears (hard tissue removed). Those materials should be prepared by curing and cooking.
3. Finely chop onion and fry in lard until golden.
4. Grinding meat.
a. Cut I, II and III beef into 3/4 - 1 inch (20 - 25 mm) pieces. You can use grinder with the kidney plate.
b. Grind skins and other meats through 1/8" (3 mm) plate.

5. Mix all meats with spices well together until mixture becomes sticky.

6. Fill the jars leaving 1 inch headspace.

7. Process at 250° F, 121° C:

1/2 Pint jars (236 ml) - 50 min.
Pint jars (473 ml) - 60 min.

Note: Original processing times for metric jars were:

350 ml jar - 50 min.
500 ml jar - 60 min.

The final product consists of pieces of meat bound together with natural gelatin. When kept cold, it should remain in one piece. The product is covered with natural gelatin (aspic, meat jelly). The might be some melted fat on the top and the bottom.

Metal Cans

Continue after step 5.

6. Fill the cans leaving 1/2 inch headspace.

7. Exhaust cans to 170° F, 77° C meat temperature.

8. Fill the cans with hot boiling water to 1/4 inch from top.

9. Process at once at 250° F, 121° C:

No. 2.5 can, 401 x 411 - 90 min.
307 x 200.25 - 45 min.

BEEF WITH BUCKWHEAT GROATS

Materials

Beef, 530 g (1.16 lb)
Buckwheat grouts (dry), 350 g (12.3 oz)
Jowl (or bacon) without the skin, 100 g (3.5 oz)
Lard, 2 g (1/2 tsp)

Beef or pork bones for making stock, 300 g (0.66 lb)

Ingredients for 1 kg of material

Salt, 15 g (2.5 tsp)
Pepper, 1.2 g (1/2 tsp)
Allspice, 1 berry
Bay leaf, one leaf,
Fresh parsley, 10 g (0.3 oz)
Fresh celery, 20 g (0.7 oz)
Fresh onion, 30 g (1 oz)

Instructions

1. Cut beef into 1 inch (25 mm) pieces. Cut jowl (bacon) into 1/2" (12 mm) pieces or grind through 1/2" (12 mm) plate.
2. Cut onion into thin discs and fry in lard until golden.
3. Place bones in 600 ml (20 oz) water and simmer at 194-203° F, 90-95° C for 3 hours. Add allspice, bay leaf, celery and parsley. Simmer additional 60 minutes. Strain the stock.
4. Wash buckwheat grouts. Place ground jowl/bacon in a pot and heat until there is some fat. Add buckwheat grouts, and 200 ml (7.76 oz) bone stock and slow cook for 60 minutes, stirring often. At the end, add fried onions and mix all together.
5. Add meat pieces to the pot and mix all together.
6. Pack 120 g (4.23 oz) of the product into the can and pour 80 ml (2.71 oz) bone stock, leaving 1/2 headspace.
7. Exhaust cans to 170° F, 77° C meat temperature.
8. Seal and process at once at 250° F, 121° C:

307 x 200.25 - 45 min.

BIGOS (HUNTER'S STEW)

Materials

Sauerkraut, 740 g (1.63 lb)
Lean pork, 80 g (2.8 oz)
Lean beef, 70 g (2.4 oz)
Smoked bacon (skinless and boneless), 50 g (1.7 oz)
Polish Smoked Sausage, 60 g (2.1 oz)

Additional Materials

Pork ribs, smoked, 70 g (2.4 oz)
Pork ribs, 70 g (2.4 oz)
Pork bones, 160 g (5.6 oz)
Lard, 3 g (1/2 tsp)

Ingredients for 1 kg of material

Salt, 35 g (6 tsp)
Pepper, 1 g (1/2 tsp)
Caraway seed, 1 g (1/2 tsp)
Bay leaf, one leaf, finely crushed
Dry mushrooms, 8 g (0.28 oz)
Onion, 8 g (0.28 oz)

Material preparation

a. Strain sauerkraut and remove about 30% of the juice in relation to the original weight. Keep the juice in case the bigos is not sauerly enogh. The juice can be added during cooking cabbage.

b. if you find your sauerkraut too acidic, rinse it with water and strain.

c. Cut the skinless cured bacon into 2 cm x 2.5 cm cm strips, dry and smoke with warm smoke for 4 hours. Cool bacon and cut into 3 mm (1/8") diameter strips.

d. Rinse dry mushrooms 3-4 times in cold water, then soak them for 30 min in cold water. Remove, drain and cut into 2 mm (1/8") strips.

e. Wash bones in cold water.

f. Cut beef and pork into 20 mm (3/4") cubes. Cut the sausage into 3 mm (1/8") discs. Cut onion into 5 mm (1/4") cubes. If using larger amounts of materials use a grinder with an appropriate plate.

Instructions

1. Place bones and ribs in a skillet and add enough water to cover. Bring to a boil, reduce heat, cover with a lid and simmer until the meat is semi-soft. Remove the ribs and separate the meat from the bones.

Place bones back in the skillet and simmer for additional 3 hours. Remove bone and remove all meat. Filter the stock through a fine sieve. Cut the rib meat into 30 mm (3/4") pieces.

2. Place cabbage in a pot and add equal by weight amount of the meat stock. Add caraway, bay leaf and cut up mushrooms and simmer everything together for 4 hours. Do not cover with the lid. Replace the meat stock lost during evaporation with more stock.

3. Melt 1/2 lard in a frying pan, then fry cut up onion just to soften it. Add beef and pork cubes and simmer for 30 minutes. Add bacon and sausage and simmer for additional 20 minutes. Stir often to prevent burning.

4. Add the following to the cooked cabbage: browned beef and pork, onion, salt, all boiled meat that was separated from the bones, and simmer for 60 minutes, stirring often. The stew should not be covered with liquid sauce, however, the sauce should come up when the stew is pressed with the spoon. Let it cool enough to be able to handle it. Add pepper and remix the stew.

5. Fill 307 x 200.25 mm cans with hot stew leaving 1/4 inch headspace.

6. Process at once 240° F, 116° C for 60 min.

BIGOS (HUNTER'S STEW) WITH TOMATOES

Materials

Sauerkraut (strained, no juice), 740 g (1.63 lb)
Lean pork, 80 g (2.8 oz)
Lean beef, 70 g (2.4 oz)
Smoked bacon (skinless and boneless), 50 g (1.7 oz)
Polish Smoked Sausage, 60 g (2.1 oz)

Additional Materials

Pork ribs, 150 g (5.3 oz)
Pork bones, 160 g (5.6 oz)
Lard, 3 g (1/2 tsp)

Ingredients for 1 kg of material

Salt, 35 g (6 tsp)
Pepper, 1 g (1/2 tsp)
Caraway seed, 1 g (1/2 tsp)
Bay leaf, one leaf, finely crushed
Dry mushrooms, 8 g (0.28 oz)
Onion, 8 g (0.28 oz)
Tomato sauce, 225 g (8 oz)

Material preparation

a. Strain sauerkraut and remove about 30% of the juice in relation to the original weight. Keep the juice in case the bigos is not sourly enough. The juice can be added during cooking cabbage.
b. if you find your sauerkraut too acidic, rinse it with water and strain.
c. Cut the skinless cured bacon into 2 cm x 2.5 cm cm strips, dry and smoke with warm smoke for 4 hours. Cool bacon and cut into 3 mm (1/8") diameter strips.
d. Rinse dry mushrooms 3-4 times in cold water, then soak them for 30 min in cold water. Remove, drain and cut into 2 mm (1/8") strips.
e. Wash bones in cold water.
f. Cut beef and pork into 20 mm (3/4") cubes. Cut the sausage into 3 mm (1/8") discs. Cut onion into 5 mm (1/4") cubes. If using larger amounts of materials use a grinder with an appropriate plate.

Instructions

1. Place bones and ribs in a skillet and add enough water to cover them. Bring to a boil, reduce heat, cover with a lid and simmer until the meat is semi-soft. Remove the ribs and separate the meat from the bones. Place bones back in the skillet and simmer for additional 3 hours. Remove bone and remove all meat. Filter the stock through a fine sieve. Cut the rib meat into 30 mm (3/4") pieces.

2. Place cabbage in a pot and add equal by weight amount of the meat stock. Add caraway, bay leaf and cut up mushrooms and simmer everything together for 4 hours. Do not cover with the lid. Replace the meat stock lost during evaporation with more meat stock.

3. Melt 1/2 lard in a frying pan, then fry cut up onion just to soften it. Add beef and pork cubes and simmer for 30 minutes. Add bacon, sausage, tomato sauce and simmer for additional 20 minutes. Stir often to prevent burning.

4. Add the following to the cooked cabbage: browned beef and pork, onion, salt, all boiled meat that was separated from the bones, and simmer for 60 minutes, stirring often. The stew should not be covered with liquid sauce, however, the sauce should come up when the stew is pressed with the spoon. Let it cool enough to be able to handle it. Sprinkle pepper and remix the stew.

5. Fill 307 x 200.25 mm cans with hot stew leaving 1/4 inch headspace.

6. Process at 240° F, 116° C for 60 min.

GOULASH

Materials

Pork grade I, lean, 600 g (1.32 lb)
Pork grade II, some fat (pork butt), 200 g (0.44 lb)
Pork grade III, connective tissue desired, 150 g (0.33 lb)
Pork skins, 50 g (0.11 lb)

Ingredients for 1 kg (2.2 lb) of material

Salt, 22 g (3 tsp)
White pepper, 1.0 g (1/2 tsp)
Cure # 1, 2.5 g (1/2 tsp)

Instructions

1. Curing mixture. Mix salt with cure #1. Allocate the mixture to each meat group: Grade I - 1.5 tsp, Grade II - 1 tsp, Grade III - 1 tsp.

2. Grinding and curing.
a. Cut pork grade I into 1" (25 mm) pieces and mix with 1.5 tsp. curing mixture.
b. Grind pork grade II with 1/4" (8 mm) plate adding 1 tsp of curing mixture.
c. Cut pork grade III into 1" (25 mm) pieces and mix with 1.5 tsp. curing mixture.
Pack tight each meat group into a different container , cover with cloth and place in refrigerator for 48 hours.

3. Skins.
a. Wash raw skins in lukewarm water.
b. Cure skins for 48 hours in 18° Be (70° salinometer, 18% salt) at refrigerator temperature. Use enough solution to be able to mix the skins without difficulty.
70° SAL solution: 1.88 lb salt to 1 gallon of water.
c. Simmer the skins for 25 minutes at 194-203° F, 90-95° C, but don't make them too soft.
d. Cool the skins on screens as warm skins stick together during grinding.
e. grind skins through 1/8" (3 mm) plate.

4. Mix ground skins with pork grade III meat, then grind through 1/8 mm (3 mm) plate. Start mixing (kneading), adding slowly 40 ml (1-1/3 oz) cold water, until mixture becomes sticky. Add remaining meat (grade I and II) and pepper and remix everything well together.

5. Fill the cans leaving 1/2 inch headspace.

6. Exhaust cans to 170° F, 77° C meat temperature.

7. Fill cans with hot boiling water to 1/4 inch from top.

8. Process at once at 250° F, 121° C:

1/2 lb, 307 x 200.25 - 45 min.
No. 2 - 307 x 409 - 60 min.
No 2.5, 401 x 411 - 70 min.

GOULASH SUPREME

Materials

Pork grade I, lean, 360 g (0.80 lb)
Pork grade II, some fat (pork butt), 450 g (1.00 lb)
Pork grade III, connective tissue desired, 100 g (0.22 lb)
Pork skins, 90 g (0.20 lb)

Additional Materials

Wheat flour, 40 g (0.14 oz)

Pork lard, 7 g (1/4 oz)

Ingredients for 1 kg (2.2 lb) of material

Salt, 15 g (2-1/2 tsp)
Pepper, 1.0 g (1/2 tsp)
Paprika, 0.5 g (1/2 tsp)
Allspice, 0.5 g (1/4 tsp)
Bay leaf, 1/2
Onion, 10 g (1/3 oz)

Instructions

1. Cut pork into 1" (25 mm) pieces.
2. Simmer the skins with bay leaf for 25 minutes at 194-203° F, 90-95° C, until semi-soft, then strain and place on screen/table to drain and cool. Save the liquid.
3. Grind the skins through 1/8" (3 mm) plate.
4. Making the roux. Heat the lard in a frying pan, adding flour. Stir continuously until *light* brown. Add 160 ml (2/3 cup) of cold skin stock (leftover liquid from boiling skins) and mix together.
5. Mix diced meats, ground skins, diced onion, spices and the roux together.
6. Pack the cans, leaving 1/2" headspace.
7. Exhaust cans to 170° F, 77° C meat temperature and remove air bubbles.
8. Fill the cans with hot boiling skin stock to 1/4" from top.
9. Process at once at 240° F, 116° C:

No. 1 (picnic), 211 x 400 - 55 min.

lAMB WITH BEANS   -

Materials

Mutton, 640 g (22.57 oz)
White beans, dry, 260 g (9.17 oz)
Pork lard, 10 g (0.35 oz)
Pork bones for making stock, 200 g (7 oz)

Ingredients for 1 kg (2.2 lb) of material

Salt, 13 g (2-1/2 tsp)
Pepper, 0.5 g (1/4 tsp)
Garlic, 3.0 (1 clove)
Paprika, 1.5 g (2 tsp)
Vinegar, 18 ml (1 Tbsp)
Wheat flour, 30 g

Instructions

1. Soak beans for 12 hours in water. Stir. Dry beans must gain 85% in weight (use scale). If more weight gain is needed soak beans longer.
2. Cut mutton into 1 inch (25 mm) pieces. Mix with salt, pepper and garlic.
3. Place pork bones in pot, cover with 600 ml (20 oz) water and boil for 4 hours. Strain the stock.
4. Making the roux. Heat the lard in a frying pan, adding flour. Stir continuously until *light* brown. Add 160 ml (2/3 cup) of pork stock (leftover liquid from boiling bones) and mix together. Add vinegar and mix.
5. Mix diced meats, ground skins, diced onion, spices and the roux together.
6 a. Pack about 160 g meat (5.64 oz) and cover with roux-stock liquid. Add about 200 g (7.05 oz) beans into pint jars. Fill with hot boiling stock leaving 1 inch headspace.
6 b. Pack about 100 g of meat (3.52 oz) and cover with roux-stock liquid. Add about 120 g (4.23 oz) of beans. Fill with hot boiling stock leaving 1 inch headspace.
7. Remove air bubbles, adjust lids and process at once at 250° F, 121° C:
1/2 Pint jars - 50 min
Pint jars - 60 min.

LAMB WITH RICE

Materials

Mutton, 670 g (23.63 oz)
Rice, 290 g (10.22 oz)
Pork lard, 40 g (1.41 oz)

Pork bones for making stock, 200 g (7 oz)

Ingredients for 1 kg (2.2 lb) of material

Salt, 15 g (2-1/2 tsp)
Pepper, 0.5 g (1/4 tsp)
Garlic, 3.0 (1 clove)
Paprika, 1.0 g (1/2 tsp)
Tomato sauce, 40 g (1.41 oz)

Instructions

1. Cut mutton into 1 inch (25 mm) pieces.
2. Fry meat pieces in lard until medium cooked.
3. Remove meat pieces and mix with salt, spices and tomato sauce. Save the meat sauce.
4. Place pork bones in pot, cover with 1 liter (1 quart) of water and boil for 4 hours. Strain the stock. Add meat sauce (from frying meat) and mix with pork stock.
5. Wash the raw rice.
6. Pack about 140 g meat (5 oz) and 100 g (3.5 oz) of rice into pint jars. Fill with hot boiling stock leaving 1 inch headspace. Pack about 85 g of meat (3 oz) and 50 g (1.76 oz) of rice. Fill with hot boiling stock leaving 1 inch headspace.
7. Remove air bubbles, adjust lids and process at once at 150° F, 121° C:

1/2 Pint jars - 40 min.
Pint jars - 50 min.

LUNCHEON MEAT

Materials

Pork grade I (lean), 650 g (1.43 lb)
Pork grade III (connective tissue), 350 g (0.77 lb)
Use pork butts for obtaining above meat grades.

Ingredients for 1 kg (2.2 lb) of material

Salt, 20 g (3 tsp)
Cure # 1, 2.5 g (1/2 tsp)
Potato flour*, 30 g (1 oz)
Water, 30 ml

Instructions

1. Curing mix. Mix salt with cure #1.
2. Cut all meat into 1 inch pieces, but keep both meat grades in separate containers.
3. Mix pork grade I with ⅔ of curing mix. Mix pork grade III with ⅓ of curing mix. Pack tightly in separate containers and cover meat with a clean cloth.
4. Cure for 72 hours in refrigerator.
5. Grind pork grade III through ⅛" (3 mm) plate.
6. Mix (knead) pork grade I until it becomes sticky. During mixing start adding potato flour. When meat is thoroughly mixed with the flour, add 30 ml of cold water and remix all together.
7. Fill cans leaving ¼ inch headspace.
8. Exhaust cans to 170° F, 77° C meat temperature.
9. Process at once at 250° F, 121° C:

No. 2, 307 x 409 - 80 min.

Note: * Potato flour binds water well. It also inhibits jelly formation which children usually don't like to eat.

LARD

Lard has always been an important cooking and baking staple in cultures where pork is an important dietary item, the fat of pigs often being as valuable a product as their meat. During the 19th century, lard was used in a similar fashion as butter in North America and many European nations. Lard was also held at the same level of popularity as butter in the early 20th century and was widely used as a substitute for butter during World War II.

Toward the late 20th century, lard began to be regarded as less healthy than vegetable oils (such as olive and sunflower oil) because of its high saturated fatty acid and cholesterol content. Despite its reputation, *lard has less saturated fat, more unsaturated fat, and less cholesterol than an equal amount of butter by weight.*

Unlike many margarines and vegetable shortenings, unhydrogenated lard contains no trans fat (the consumption of trans fats increases the risk of coronary heart disease). It has also been regarded as a "poverty food". Many restaurants in the western nations have eliminated the use of lard in their kitchens *because of the religious and health-related dietary restrictions of many of their customers.* Many industrial bakers substitute beef tallow for lard in order to compensate for the lack of mouthfeel in many baked goods and to free their food products from pork-based dietary restrictions.

In the 1990's and early 2000's, the unique culinary properties of lard became widely recognized by chefs and bakers, leading to a partial rehabilitation of this fat among professionals. This trend has been partially driven by negative publicity about the trans fat content of the partially hydrogenated vegetable oils in vegetable shortening. Lard has become popular in the United Kingdom among aficionados of traditional British cuisine. This led to a "lard crisis" in early 2006 in which British demand for lard was not met due to demand by Poland and Hungary (who had recently joined the European Union) for fatty cuts of pork that had served as an important source of lard. Lard is one of the few edible oils with a relatively high smoke point, attributable to its high saturated fatty acids content. Pure lard is especially useful for cooking since *it produces little smoke when heated* and has a distinct taste when combined with other foods. Many chefs and bakers deem lard a superior cooking fat over shortening because of lard's range of applications and taste. Because of the relatively large fat crystals found in lard, it is extremely effective as a shortening in baking.

Pie crusts made with lard tend to be more flaky than those made with butter. Many cooks employ both types of fat in their pastries to combine the shortening properties of lard with the flavor of butter.

Butter consists mostly of *saturated fat* and is a significant source of dietary cholesterol. For these reasons, butter has been generally considered to be a contributor to health problems, especially heart disease. For many years, vegetable margarine was recommended as a substitute, since it is an unsaturated fat and contains little or no cholesterol. In recent decades, though, it has become accepted that the trans fats contained in partially hydrogenated oils used in typical margarines significantly raise undesirable LDL cholesterol levels as well.

Comparative properties of common cooking fats per 100 g (3.5 oz)						
	Total Fat	Saturated Fat	Mono-saturated Fat	Polyun-saturated Fat	Protein	Choles-terol
Vegetable shortening	71 g	23 g	8 g	37 g	0 g	0
Olive Oil	100 g	14 g	73 g	11 g	0 g	0
Butter	81 g	51 g	21 g	3 g	1 g	215 mg
Tallow	100 g	50 g	42 g	4 g	0 g	109 mg
Lard	100 g	39 g	45 g	11 g	0 g	95 mg
Lard is rendered (heat melted) pork fat. *Tallow is a rendered suet.* *Suet is a fresh beef or sheep fat.*				*Source: USDA Nutrient database*		

A higher proportion of monounsaturated fats in the diet is linked with a reduction in the risk of coronary heart disease. This is significant because olive oil is considerably rich in monounsaturated fats, most notably oleic acid. *Lard contains more monounsaturated fats than butter or tallow*. Lard is one of the great products that is largely misunderstood today and has developed an undeserving reputation as an unhealthy product. To set the record straight we have listed some USDA data and statistics about lard and other animal fats.

The conclusion is simple: *pork fat (lard) is much healthier than butter or tallow (beef or sheep fat)*. Save on butter, eat more lard and you will live longer.

When the second World War ended in 1945, it was the main staple in every household for these reasons:

- 100 g of lard provides 900 kcal of energy.
- It tasted great. People ate lard sandwiches with tomatoes, pickles, it tasted even great when topped with sugar.
- Could be kept in a cool place for many months.

Home Production of Lard

Lard can be obtained from any part of the pig as long as there is a high concentration of fatty tissue. The highest grade of lard, known as leaf lard, is obtained from the "flare" visceral fat deposit surrounding the kidneys and inside the loin. Leaf lard has *little pork flavor*, making it ideal for use in baked goods, where it is treasured for its ability to produce flaky, moist pie crusts. The next highest grade of lard is obtained from *fatback*, the hard subcutaneous fat between the back skin and muscle of the pig. To extend the shelf life of pork fat (or any fat) it must be rendered (melted down). This offers the following benefits:

- Removal of water.
- Separation of impurities.
- It kills bacteria that would start the spoiling process.

Photo 23.6 Home made lard.

Photo 23.7 Lard on bread.

Two types of lard can be produced:

1. Lard shortening that will be used for general cooking and frying. Such a lard can be made from any pork fat and even 20% of beef fat may be added. The leftover cracklings are normally saved for making liver or blood sausages. The easiest way to make lard is to mince fat with a grinder and that will produce the largest amount of lard.

Instructions:

- Cut fat into 1" pieces and grind through ⅛" (3 mm) plate.
- Add little water (½ cup to 4 quart skillet) to a skillet and place on a stove. Adding water prevents lard from sticking to the bottom of a skillet. Lard being a fat, will not mix with water anyhow and the water will evaporate during cooking.
- Add ground fat and stir often in order not to burn the fat.
- A by-product of dry-rendering lard is deep-fried meat, skin and membrane tissue known as cracklings. Once when cracklings develop a golden color and the lard becomes clearer, take off the skillet from the stove and let it stand for 20 minutes.
- Pour 75% of clear lard into jars. This is very clean lard that will last the longest. Filter the remaining lard (it contains cracklings) through a fine sieve or cheese cloth. Use cracklings for sausages. Use this lard first.

2. Ready to eat lard (*smalec* in Polish) has been traditionally produced in Europe to be spread on a slice of bread and eaten as a sandwich. Such lard is made *from pork fat only*, preferably from back fat. Belly fat may be used as well although it may be considered a waste as bacon can be processed in many other ways. The resulting cracklings are saved and become a part of lard. They may be added to boiled sausages (liver and blood). *The highest quality lard* will be obtained when the pork fat is manually diced into ¼" cubes which will produce a larger number of solid cracklings, known in Polish as *skwarki*. To add extra flavor, ingredients such as onion, garlic, apple or marjoram are often added.

Instructions:

- Cut fat into 1" pieces and grind through ⅛" (3 mm) plate.
- Add little water (½ cup to 4 quart skillet) to a skillet and place on a stove.
- Add ground fat and stir often in order not to burn the fat.
- Once when cracklings develop a golden color and the lard becomes clearer, take off the skillet from the stove and let it stand for 20 minutes.
- Pour 75% of clear lard into jars. This is very clean lard that will last longest. Filter the remaining lard (it contains cracklings) through a fine sieve or cheese cloth. Use this lard first.
- Add cracklings to each container or a glass jar. They will have a tendency to settle down on the bottom. To distribute them evenly, mix lard adding a new portion of cracklings. As the lard cools down it changes color to white. The cracklings will be trapped inside and uniformly distributed.

Adding Flavors

Follow the above procedure and when the lard is half-way done (becomes semi-liquid), add chopped onion. Add onion carefully as it contains water and the lard may boil over. Then continue as usual. When chopped onion is added to lard, the lard should be refrigerated. If lard will be kept at room temperature, add whole peeled onion and then discard the onion during filtering. If not kept under refrigeration, chopped onion will decrease the shelf life of the product. This is why it is added to canned products in such a small amount. Add whole onions, a few cloves of peeled garlic, apples cored and cut in halves, and a bay leaf. Remove those ingredients during filtering. Add spices of your choice: marjoram, coriander or others. Lard consumed as a spread on bread was once very common in Europe and North America, especially those areas where dairy fats and vegetable oils were rare.

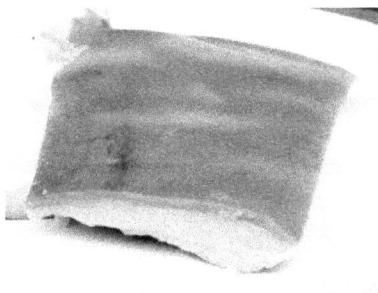

Photo 23.8 Modern back fat contains little fat.

Photo 23.9 In the past back fats were about 2" thick.

Photo 23.10 Diced back fat.

Photo 23.11 Melting fat.

Photo 23.12 Filtered lard.

Photo 23.13 Cracklings. May be added to lard or used in blood sausages.

LARD - HOMEMADE

Materials

Back fat, fresh (not frozen), without skin and traces of meat, 1 kg (2.2 lb)

Instructions

1. Dice back fat into 1/4 inch cubes or grind through a 3/8" plate.
2. Put skillet on a heat source and add diced/ground back fat. Apply heat and start melting the fat, stirring often.
Any indication of smoke signifies that the temperature is too high, turn the heat down.
3. Remove cracklings by filtering the lard through a strainer.
4. When the temperature of the lard drops to 72° F, 22° C, pour the lard into No. 2.5, 401 x 411 or No. 2, 307 x 409 cans.
Place in a cool area so the lard will set.
If needed, you can add more warm (liquid) lard to the container.
5. Seal the cans.
6. Clean the cans and store.

Note: there is no thermal processing involved.

There were very few products that were not submitted to thermal process. The products were lard and smoked bacon, both packed in cans. Smoked sausages were often packed in lard to extend their shelf life. The practice was not limited to Eastern European countries, but was common in other areas, for example in Cuba. Such products were kept in root cellars or in cool pantries.

Fresh fat is a poor medium for bacteria to grow as it contains very few nutrients and little moisture (15%) as compared to meat which contains proteins, vitamins, minerals and 75% moisture. During production of lard most of the moisture is evaporated during heating and this is the reason lard keeps so well.

LARD - PLAIN

Materials

Back fat, fresh (not frozen) , without skin and traces of meat, 1 kg
(2.2 lb)
Salt 15 g (2.5 tsp)

Instructions

1. Grind back fat through 1/4" (5 mm) plate adding salt.
2. Mix the ground fat to distribute uniformly salt.
3. Warm up the fat slightly (100° F, 38° C) then pour into warm jars
leaving 1 inch headspace.
4. Process without delay at 250° F, 121° C:

1/2 pint jar - 50 min.
Pint jar - 60 min.

Metal Cans

Follow the above instructions. Exhausting and sealing up hot lard
is not recommended for home production due to safety risks when
handling hot oil.

Fill the cans leaving 1/4 inch headspace.

Process at once at 250° F, 121° C:
1/2 lb, 307 x 200.25 - 60 min.

LARD -SMOKED-WITH SPICES

Materials

Back fat, fresh (not frozen), without skin and traces of meat, 900 g (1.98 lb)
Smoked back fat - no older than 4 weeks from the date it was smoked, 100 g (0.22 lb)
Salt, 15 g (2.5 tsp)
Garlic, 1.0 g (1/3 clove)
Marjoram, dry, 0.5 g (1/2 tsp)

Instructions

1. Dice fresh garlic finely, then mix/grind with a small amount of salt. Use a spoon and the cup or mortar and pestle for that purpose.
2. Grind fresh back fat and smoked back fat together through 1/4" (5-6 mm) plate adding all ingredients.
3. Remix ground fat to distribute uniformly the ingredients.
4. Warm up the fat slightly (100° F, 38° C) then pour into warm jars leaving 1 inch headspace.
5. Process without delay at 250° F, 121° C:

1/2 pint jar - 50 min.
Pint jar - 60 min.

Metal Cans

Follow the above instructions. Exhausting and sealing up hot lard is not recommended for home production due to safety risk when handling hot oil.

Fill the cans leaving 1/4 inch headspace.

Process at once at 250° F, 121° C:

1/2 lb, 307 x 200.25 - 60 min.

LARD WITH ONIONS

Materials

Back fat, fresh (not frozen), without skin and traces of meat, 1.0 kg (2.2 lb)
Salt, 15 g (2.5 tsp)
Onion, 10 g (1/3 oz)

Instructions

1. Peel the onions, slice into discs and fry in fat until golden (don't brown them).
2. Grind back fat through 1/4" (5-6 mm) plate together with salt and onion.
3. Mix the ground fat to distribute uniformly the ingredients.
4. Warm up the fat slightly (100° F, 38° C) then pour into warm jars leaving 1 inch headspace.
5. Process without delay at 250° F, 121° C:

1/2 pint jar - 50 min.
Pint jar - 60 min.

Metal Cans

Follow the above instructions. Exhausting and sealing up hot lard is not recommended for home production due to safety risk when handling hot oil.

Fill the cans leaving 1/4 inch headspace.

Process at once at 250° F, 121° C:

1/2 lb, 307 x 200.25 - 60 min.

PATE POPULAR

Materials

Pork, I, II and III grade, 550 g
Pork skins and pork feet meat, 100 g
Liver, pork or veal, 100 g
Fat trimmings, 250 g

Ingredients for 1 kg (2.2 lb) of material

Salt, 12 g (2 tsp)
Pepper, 1 g (0.5 tsp)
Allspice, 0.3 g (1/4 tsp)
Marjoram, 0.2 g
Bay leaf, crushed, 1/2
Onion, 20 g (1/2 small onion)

Instructions

1. Wash the liver, cut it across and then scald in hot water (167° F, 75° C) until no more blood is released.
2. Wash the skins and feet meat. Simmer in water (194° F, 90° C) for two hours.
3. Simmer all pork and fat trimmings with 1/2 bay leaf in a little water (194° F, 90° C) for about 30 minutes.
4. Drain and leave on the screen/table to cool down.
5. Cut the onion in discs and fry in fat until golden.
6. Grind liver, meat, fat trimmings and onion through 1/8" (3 mm) plate. Add all spices and either grind again or process in food processor, until one smooth paste is obtained.
7. Pack into glass jars leaving 1 inch headspace.
8. Place jars into warm water and heat up to about 100° F, 38° C.
9. Seal jars, and process at once at 250° F, 121° C:

1/2 Pint - 50 min.
Pint - 60 min.

PATE SUPREME

Materials

Pork liver, 250 g (0.55 lb)
Pork grade II, some fat (pork butt), 400 g (0.88 lb)
Pork trimmings, fat and connective tissue permitted,50 g (2 oz)
Fat trimmings, 300 g (0.66 lb)

Ingredients for 1 kg (2.2 lb) of material

Salt, 14 g (2.5 tsp)
Pepper, 1.5 g (0.75 tsp)
Allspice, 0.3 g (1/4 tsp)
Nutmeg, 0.5 g (1/4 tsp)
Bay leaf, one leaf, finely crushed
Cloves, 0.2 g (1/10 tsp)
Onion, 10 g (1/4 small onion)

Instructions

1. Wash the liver, cut it across and then scald it in hot water (167° F, 75° C) until no more blood is released.
2. Dice onion, take 10 g of fat trimmings and fry onion until golden.
3. Precook all meats and fat trimmings (except liver) in a skillet, stirring often, until meat looses its raw meat color.
4. Grind liver, meat and fat trimmings through 1/8" (3 mm) plate or smaller. Add all spices and either grind again or process in food processor, until one smooth paste is obtained.
5. Pack into glass jars leaving 1 inch headspace.
6. Place jars into warm water and heat up to about 100° F, 38° C.
7. Seal jars, and process at once at 250° F, 121° C:

1/2 Pint - 50 min
Pint - 60 min

Metal Cans

Continue after step 4:
5. Fill the cans leaving 1/4 inch headspace.
6. Exhaust cans to 170° F, 77° C meat temperature.
7. Seal and process at once at 250° F, 121° C:
307 x 200.25 - 50 min.
307 x 409 - 80 min.

PORK AND VEAL

Materials

Pork grade II (with fat), 510 g (18 oz)
Lean veal, 250 g (8.81 oz)
Regular veal (with fat), 150 g (5.29 oz)
Pork skins, 80 g (2.82 oz)
Pork lard, 10 g (0.35 oz)

Ingredients for 1 kg (2.2 lb) materials

Salt, 18 g (3 tsp)
Pepper, 0.5 g (1/4 tsp)
Allspice, 0.2 g (1/10 tsp)
Marjoram, 0.2 g (1/5 tsp)
Small onion, 1/2, 20 g (0.70 oz)

Instructions

1. Skins.
a. Wash raw skins in a lukewarm water.
b. Cure skins for 48 hours in 18° Be (70° salinometer, 18% salt) at refrigerator temperature. Use enough solution to be able to mix the skins without difficulty.
70° SAL solution: 1.88 lb salt to 1 gallon of water.
c. Simmer the skins for 25 minutes at 194-203° F, 90-95° C, but don't make them too soft.
d. Cool the skins on screens as warm skins glue together during grinding.
2. Dice onion and fry in lard until golden.
3. Cut pork and veal into 1" (25 mm) pieces.
4. Grind skins through 1/8" (3 mm) plate.
5. Mix meats with ground skins adding salt, until the mixture becomes sticky. Add onion and spices and remix.
6. Pack the cans leaving 1/2 inch headspace.
7. Exhaust cans to 170° F, 77° C meat temperature.
8. Add hot boiling water (if needed) 1/4 inch to top.
9. Process at once at 250° F, 121° C:

307 x 200.25 - 50 min.
307 x 409 - 70 min.
No. 2.5, 401 x 411 - 80 min.

PORK - GROUND

Materials

Pork grade I (lean), 300 g (0.66 lb)
Pork grade II (with fat), 500 g (1.10 lb)
Pork grade III (with connective tissue), 200 g (0.44 lb)

Ingredients for 1 kg (2.2 lb) of material

Salt, 20 g (3-1/2 tsp)
Cure #1, 2.5 g (1/2 tsp)
Pepper, 1 g (0.5 tsp)
Potato flour, 50 g (1.76 oz)

Instructions

1. Dry cure mix. Mix salt with cure #1.
2. Cut pork grade I into 1 inch pieces. Add 1/3 of dry cure mix during grinding.
Grind pork grade II through 1/2" (12 mm) plate. Add 1/3 of dry cure mix during grinding.
Grind pork grade III through 1/8 plate (3 mm). Add 1/3 tsp of dry cure mix during grinding. Re-grind again adding 6 ml (1 tsp) of water.
3. Mix all meats together adding pepper. When the mixture feels sticky add gradually potato flour and keep on mixing.
4. Pack cans with about 425 g (15 oz) leaving 1/4 inch headspace.
5. Exhaust cans to 170° F, 77° C meat temperature.
6. Process at once at 240° F, 116° C:

301 x 408 - 55 min.

PORK HOCKS IN ASPIC

Materials

Pork hocks 1000 g (2.2 lb)

Ingredients for 1 kg of materials

Salt for curing, 80 g (2.82 oz)
Cure # 1,
Mustard seeds, 0.5 g (1/4 tsp)
Bay leaf, 1
Maggi liquid seasoning, 5 g (1 tsp)
Gelatin*, 5 g

* Instead of commercial gelatin a natural gelatin can be obtained by making strong stock from pig feet or skins.

Instructions

1. Curing solution 14 Be, 55° SAL. Dissolve 1.15 lb (523 g) of salt in 1 gallon of cold water. Add 120 g (4.2 oz) of cure #1 (there is 112 g of salt in cure #1 and it is included in the calculation).
2. Remove bones from hocks using as few cuts as possible. Using meat pump inject each meat cut with curing solution. Place pumped meats in a container and fill with curing solution. Use 60% solution to 40% of meat by weight, for example: 400 g of meat and 600 ml solution. Cure for 48-72 hours in refrigerator. Place for 8 hours on screens to drain at refrigerator temperature (40° F, 4° C).
3. Simmer meats at 194-203° F, 90-95° C for 30 minutes. Cut meat to smaller pieces, about 120 g (4 oz).
4. Dissolve commercial gelatin in cold water, then bring to a boil adding bay leaf and Maggi liquid seasoning. Simmer for 30 minutes.
5. Pack 2-3 pieces of meat, about 400 g (14 oz) total weight into pint jar, add a few mustard seeds and a piece of bay leaf. Pack 1-2 pieces of meat, about 200 g (7 oz) into 1/2 pint jar, adding a few mustard seeds and a part of bay leaf. Pour hot gelatin solution, leave 1 inch headspace, remove air bubbles, adjust lids and process at once at 250° F, 121° C:

1/2 Pint jar - 30 min.
Pint jar - 40 min.

Metal Cans

Steps 1-4 remain the same.

5. No. 2.5 can, 401 x 411: Pack 800 g meat (28.22 oz) and fill 75 ml (2.5 oz) gelatin solution. Add 12 mustard seeds.

301 x 408 can: Pack 400 g meat (14.11 oz) and fill with 30 ml (1 oz) gelatin solution. Add 6 mustard seeds.

Pack cans leaving 1/2 inch headspace.

6. Exhaust cans to 170° F, 77° C meat temperature. Remove air bubbles. Add hot gelatin solution to 1/4 from top.

7. Seal and process at once at 250° F, 121° C:

301 x 408 - 60 min.
401 x 411 - 75 min.

PORK IN ITS OWN JUICE

Materials

Pork grade I, lean, 300 g (0.66 lb)
Pork grade II, some fat (pork butt), 518 g (1.14 lb)
Pork head meat *, 100 g (0.22 lb)
Pork skins, 80 g (2.82 oz)
Lard, 2 g (1/2 tsp)

* lean meat cuts may be used instead of head meat, but head meat is more flavorsome.

Ingredients for 1 kg (2.2 lb) of material

Salt, 18 g (3.0 tsp)
Pepper, 1.0 g (1/2 tsp)
Marjoram, 0.2 g (1/10 tsp)
Onion, 10 g (1/4 small onion) or 1.5 g dry onion (1 tsp)

Instructions

1. Remove any visible fat from pork cuts.

2. Heads. If the jowl is attached to the head remove it.
a. Soak pork heads in cold water for 4 hours. Change water twice.
b. Cook heads for 30 minutes in boiling water, then lower the temperature to 194-203° F, 90-95° C and simmer the heads until the meat can be separated from bones. This will take about 90 minutes. Do not overcook the meat.
c. Place the heads on the table and cool until they can be handled.
d. Separate all meat from bones, remove the tough ear tissue.

3. Skins.
a. Wash raw skins in lukewarm water.
b. Cure skins for 2 days in 18° Be (70° salinometer, 18% salt) at refrigerator temperature. Use enough solution to be able to mix the skins without difficulty.
70° SAL solution: 1.88 lb salt to 1 gallon of water.
c. Simmer the skins for 25 minutes at 194-203° F, 90-95° C, but don't make them too soft.
d. Cool the skins on screens as warm skins glue together during grinding.

4. Finely chop the onion and fry in lard until golden.

5. Grinding meat.

a. Cut I and II pork into 3/4 - 1 inch (20 - 25 mm) pieces. You can use grinder with the kidney plate.

b. Grind head meat through 3/8" (10 mm) plate.

c. Grind skins through 1/8" (3 mm) plate or smaller. If no small plate is available, run them through a food processor.

6. Mix all meats with spices until mixture becomes sticky.

7. Fill the jars leaving 1 inch headspace.

8. Process at once at 250° F, 121° C:

1/2 Pint jars (236 ml) - 50 min.
Pint jars (473 ml) - 60 min.

Note: Original processing times for metric jars were:

350 ml jar - 50 min.
500 ml jar - 60 min.

The final product consists from pieces of meat bound together with natural gelatin. When kept cold, it should remain in one piece.

Metal Cans

Continue after step 6.

7. Fill the cans leaving 1/2 inch headspace.

8. Exhaust cans to 170° F, 77° C meat temperature.

9. Fill the cans with hot boiling water to 1/4 inch from top.

10. Process at once at 250° F, 121° C:

1/2 lb can, 307 x 200.25 - 50 min.
No. 2 can, 307 x 409 - 70 min.
No. 2.5 can, 401 x 411 - 80 min.

PORK WITH BEANS IN TOMATO SAUCE

Pork grade II (with fat, butt), 450 g (0.99 lb)
Pork grade III (connective tissue), 150 g (0.33 lb)
Pork lard, 10 g (0.35 oz)
Dry beans, 390 g (0.85 lb)

Ingredients for 1 kg (2.2 lb) of material

Salt, 18 g 3 tsp)	Maggi Sauce*, 2 ml (1/2 tsp)
Pepper, 0.5 g (1/4 tsp)	Vinegar, 5 ml (1 tsp)
Paprika, 0.5 g (1/4 tsp)	Onion, small, 1/2, 20 g (0.7 oz)
Nutmeg, 0.3 g (1/8 tsp)	Wheat flour, 14 g (0.49 oz)
Allspice, 0.3 g (1/8 tsp)	Potato flour, 24 g (0.84 oz)
Sugar, 3 g (1/2 tsp)	Tomato sauce 50 g (1.76 oz)

* Maggi Sauce can be bought in every supermarket

Instructions

1. Soak beans for 12 hours in water. Stir. Dry beans must gain 80% in weight (use scale). If more weight gain is needed soak beans longer. Scald soaked beans in boiled water for 3 minutes. Drain and spread on table to cool.
2. Cut pork grade II into 1 inch (25 mm) pieces. Grind pork grade III through 1/8" (3 mm) plate.
3. Mix ground pork grade III with 6 g salt (1 tsp), then add pieces of pork grade II, 12 g salt (2 tsp) and keep on mixing until sticky. Add potato flour and remix.
4. Making roux. Heat the lard in a frying pan, adding flour. Stir continuously until *light* brown.
5. Sauce. Add to roux one pint (373 ml) of tomato sauce and mix together. Add all spices, Maggi, vinegar and bring to a boil, stirring often. Turn the heat off.
6. Pack cans with 70 g (2.47 oz) meat mass, add 80 (2.82) g soaked and scalded beans and fill with sauce leaving 1/2 inch headspace.
7. Exhaust cans to 170° F, 77° C meat temperature. Fill with hot sauce to 1/4 inch from top.
8. Seal and process at once at 250° F, 121° C:

307 x 200.25 can - 45 min.

TRIPE STEW

Materials

Beef stomachs*, 450 g (0.99 lb)
Pork stomachs*, 450 g (0.99 lb)
Pork skins, 60 g (2 oz)
Lard, 10 g (1/3 oz)
* clean tripe can usually be bought in food supermarkets
Beef bones for making stock, 1 kg (2.2 lb)

Ingredients for 1 kg (2.2 lb) of material

Salt, 12 g (2 tsp)	Parsley root, 10 g (1/3 oz)
Pepper, 1.0 g (1/2 tsp)	Carrots, 10 g (1/3 oz)
Ginger, 0.5 g (1/2 tsp)	Marjoram, 0.2 g (1/10 tsp)
Paprika, 0.5 g (1/2 tsp)	Celery, 10 g (1/3 oz)

Instructions

1. Both beef and pork stomachs are prepared in the same manner. Clean stomachs well, than soak for 12 hours in cold water. A few water changes are needed. Cur stomachs into two parts, place in water and bring to a boil. Cook for 15 minutes. Remove from water and check for a smell. If an odor is still detected, boil stomachs for additional 15 minutes and the odor will go away. Place stomachs on screens or tables to cool.
2. Wash the skins and simmer at 194-203° F, 90-95° C for 60 minutes in about 120 ml (1/2 cup) of water. Spread on table to cool down.
3. Place bones in pot, add 5 liters (5 quarts and one cup) of water and bring to a boil. Cook for 3 hours, add parsley, celery and carrots and cook one more hour (total 4 hours). Strain the stock.
4. Cut stomachs into strips 2 x 1/4" (50 x 6 mm). Grind skins through 1/8" (3 mm) plate.
5. Mix tripe (stomach strips) with ground skins adding salt, remaining spices and melted lard.
6. Place 280 g (10 oz) tripe mixture in pint jar, and 140 g (5 oz) tripe in 1/2 pint jars. Add boiling hot beef stock leaving 1 inch headspace.
7. Process at once at 250° F, 121° C:

1/2 Pint jars - 50 min.
Pint jars - 60 min.

TRIPE STEW IN BROTH

Materials

Beef tripe (stomachs)*, 930 g (2.0 lb)
Pork skins, 60 g (3 oz)
Lard, 10 g (1/3 oz)
Beef bones for making stock, 1 kg 92.2 lb)
* clean tripe can usually be bought in food supermarkets

Ingredients for 1 kg (2.2 lb) material

Salt, 12 g (2 tsp)	Ginger, 0.2 g (1/10 tsp)
Pepper, 1 g (1/2 tsp)	Parsley,fresh, 10 g (1/3 oz)
Paprika, 0.5 g (1/4 tsp)	Celery, fresh, 10 g (1/3 oz)
Marjoram, 1.0 g (1/2 tsp)	Carrots, fresh, 10 g (1/3 oz)

Instructions

1. Both beef and pork stomachs are prepared in the same manner. Clean stomachs well, than soak for 12 hours in cold water. A few water changes are needed. Cur stomachs into two parts, place in water and bring to a boil. Cook for 15 minutes. Remove from water and check for a smell. If an odor is still detected, boil stomachs for additional 15 minutes and the odor will go away. Place stomachs on screens or tables to cool.
2. Wash the skins and simmer at 194-203° F, 90-95° C for 60 minutes in about 120 ml (1/2 cup) of water. Spread on table to cool down. Grind through 1/8" (3 mm) plate or smaller.
3. Place bones in pot, add 5 liters (5 quarts and one cup) of water and bring to a boil. Cook for 3 hours, add parsley, celery and carrots and cook one more hour (total 4 hours). Strain the stock.
4. Cut stomachs into strips 2 x 1/4" (50 x 6 mm). Grind skins through 1/8" (3 mm) plate. Mix the tripe and ground skins together adding salt, remaining spices and melted lard.
5. Place 280 g (10 oz) tripe mixture in pint jar, and 140 g (5 oz) tripe in 1/2 pint jars. Add boiling hot beef stock leaving 1 inch headspace.
6. Process at once at 250° F, 121° C:
1/2 Pint jars - 40 min.
Pint jars - 60 min.

Metal Cans

Processing steps 1-4 remain the same.
5. Pack 550 g (19.40 oz) of tripe mixture in the can and pour 280 ml (9.47 oz) of beef stock, leaving 1/2 inch headspace.
6. Exhaust cans to 170° F, 77° C meat temperature. Remove air bubbles and ad hot boiling stock to 1/4 inch from top.
7. Seal and process at once at 250° F, 121° C: 401 x 411 can - 60 min.

Links of Interest

Glass Jars, Metal Cans, Pressure Canners, Can Sealers

Chef's Design
www.allamerican-chefsdesign.com

Freund Container & Supply
www.freundcontainer.com

Wells Can Company Limited
www.wellscan.ca

House of Cans, Inc.
www.houseofcans.com

Store-it Foods
www.storeitfoods.com

Embarcadero Home Cannery
www.ehcan.com

Pressure Cooker Canner Co.
www.pressurecooker-canner.com

Fillmore Container
www.fillmorecontainer.com glass containers, vacuum gauge

Canning Pantry
www.canningpantry.com

Crown Holdings, Inc.
www.crowncork.com/products_services/food_offerings_cans_round.php

Dixie Canner Co.
www.dixiecanner.com can packaging and processing equipment

Small Hand Tools And Test Instruments

WACO-Wilkens-Anderson Company
www.wacolab.com/subcategories.asp?cat=Can+Test+Apparatus

Terris Consolidated Industries, Inc. *can cover opener*
www.terriss.com/productDetail.aspx?ProductID=270#More

Can Defects And Double Seam Troubleshooting Guides

Double Seam
www.doubleseam.com/

Shoreline Packaging And Processing Machinery
http://shorelineppm.com/engineering.htm

EVCO Wholesale Food Corp.
Classification of Visible External Can Defects
www.evcofoods.com/pdf/AOAC%20Can%20Defects.pdf
A Pocket Guide to Can Deffects
www.evcofoods.com/pdf/A%20Pocket%20Guide%20To%20
Can%20Defects.pdf

FDA
Low Acid Canned Food Manufacturers: Part 3, Container Closures
www.fda.gov/ICECI/Inspections/InspectionGuides/ucm074999.htm

Canadian Food Inspection Agency
Metal Can Defects Manual - Identification and Classification
http://www.inspection.gc.ca/food/fish-and-seafood/manuals/metal-
can-defects/eng/1348848316976/1348849127902

University of Alaska in Fairbanks
Canning low-acid foods in jars and cans
www.uaf.edu/ces/preservingalaskasbounty/

Regulations

ELECTRONIC CODE OF FEDERAL REGULATIONS
www.ecfr.gov/cgi-bin/ECFR?page=browse
Canning, FDA - Title 21, Parts: 108, 113, 114
Canning, USDA - Title 9, Parts: 318, 416

FDA Regulations And Guides
Publications
www.gmppublications.com

Californian Regulations
www.cdph.ca.gov/services/Documents/fdb%20Can%20Regs%20
T17%2012400.pdf

Index

Other Books by Stanley & Adam Marianski

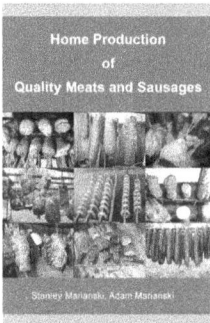

Home Production of Quality Meats And Sausages bridges the gap that exists between highly technical textbooks and the requirements of the typical hobbyist. The book covers topics such as curing and making brines, smoking meats and sausages, making special sausages such as head cheeses, blood and liver sausages, hams, bacon, butts, loins, safety and more...

ISBN: 978-0-9824267-3-9

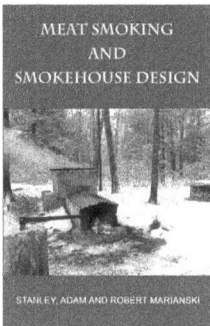

Meat Smoking & Smokehouse Design explains differences between grilling, barbecuing and smoking. There are extensive discussions of curing as well as the particulars about smoking sausages, meat, fish, poultry and wild game.

ISBN: 978-0-9824267-0-8

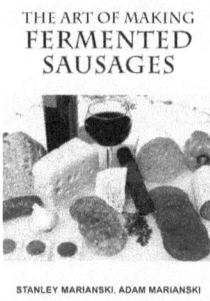

The Art of Making Fermented Sausages shows readers how to control meat acidity and removal of moisture, choose proper temperatures for fermenting, smoking and drying, understand and control fermentation process, choose proper starter cultures and make traditional or fast-fermented products, choose proper equipment, and much more...

ISBN: 978-0-9824267-1-5

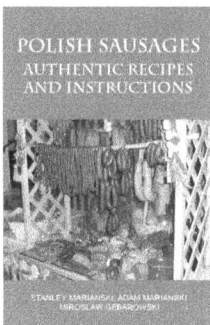

Polish Sausages contains government recipes that were used by Polish meat plants between 1950-1990. These recipes come from government manuals that were never published before, which are now revealed in great detail.

ISBN: 978-0-9824267-2-2

Sauerkraut, Kimchi, Pickles and Relishes teaches you how to lead a healthier and longer life. Most commercially produced foods are heated and that step eliminates many of the beneficial bacteria, vitamins and nutrients. However, most of the healthiest vegetables can be fermented without thermal processing. The book explains in simple terms the fermentation process, making brine, pickling and canning.

ISBN: 978-0-9836973-2-9

Making Healthy Sausages reinvents traditional sausage making by introducing a completely new way of thinking. The reader will learn how to make a product that is nutritional and healthy, yet delicious to eat. The collection of 80 recipes provides a valuable reference on the structure of reduced fat products.

ISBN: 978-0-9836973-0-5

The Amazing Mullet offers information that has been gathered through time and experience. Successful methods of catching, smoking and cooking fish are covered in great depth and numerous filleting, cleaning, cooking and smoking practices are reviewed thoroughly. In addition to mullet recipes, detailed information on making fish cakes, ceviche, spreads and sauces are also included.

ISBN: 978-0-9824267-8-4

Home Production of Vodkas, Infusions & Liqueurs is a guide for making quality alcohol beverages at home. The book adopts factory methods of making spirits but without the need for any specialized equipment. A different type of alcohol beverage can be produced from the same fruit and the authors explain in simple terms all necessary rules.

ISBN: 978-0-9836973-4-3